UNITEXT for Physics

UNITEXT for Physics series publishes textbooks in physics and astronomy, characterized by a didactic style and comprehensiveness. The books are addressed to upper-undergraduate and graduate students, but also to scientists and researchers as important resources for their education, knowledge, and teaching.

Roman Brajša · Arnold Hanslmeier

Solar Rotation

Springer

Roman Brajša
Hvar Observatory, Faculty of Geodesy
University of Zagreb
Zagreb, Croatia

Arnold Hanslmeier
Institute of Physics
University of Graz
Graz, Austria

ISSN 2198-7882 ISSN 2198-7890 (electronic)
UNITEXT for Physics
ISBN 978-981-97-6878-3 ISBN 978-981-97-6879-0 (eBook)
https://doi.org/10.1007/978-981-97-6879-0

This Springer imprint is published by the registered company Springer Nature Singapore Pte Ltd.
The registered company address is: 152 Beach Road, #21-01/04 Gateway East, Singapore 189721, Singapore

If disposing of this product, please recycle the paper.

Roman Brajša dedicates this book to his wife Karmen and to his daughter Romana. Arnold Hanslmeier dedicates this book to his partner Anita and to his children Roland, Christina, and Alina. Both authors dedicate this book to all their colleagues and friends with whom they can share the enthusiasm and passion for astronomy and astrophysics and especially for solar and stellar physics.

Preface

In the "Feynman's Epilogue" at the end of the Vol. III of the famous "The Feynman Lectures on Physics" by R. P. Feynman, R. B. Leighton, and M. Sands (*Addison-Wesley Publ. Comp., Reading, Massachusetts.* 1965) it stands:

> ... may I add that the main purpose of my teaching has not been to prepare you for some examination—it was not even to prepare you to serve industry or the military. I wanted most to give you some appreciation of the wonderful world and the physicist's way of looking at it, which, I believe, is a major part of the true culture of modern times. ... Perhaps you will not only have some appreciation of this culture; it is even possible that you may want to join in the greatest adventure that the human mind has ever begun.

We, the authors of this book, took this challenge and devoted a part of our professional research in astrophysics to the topic of solar rotation, which is presented in the book.

The Sun is our host star and the only one where the surface phenomena can be observed in detail. The first telescopic observations of the Sun were performed by G. Galilei and Ch. Scheiner and by their contemporaries in the early seventeenth century, revealing the existence of sunspots (discovered already in the pre-telescopic era) and enabling following them on the solar disk. Already then it was discovered that the sunspots at different solar latitudes rotate with different periods. This was also the discovery of the solar differential rotation.

Ever since the solar rotation has been one of the extensive topics of the solar research. During the last four centuries, the measurement techniques were steadily improved and refined. So, our knowledge of the solar rotation expanded and it was found that it shows variations with the solar latitude, the height in the solar atmosphere (and the depth below the surface) as well as with the time. Later, when the solar activity cycle and solar magnetism were discovered, a close relationship between the solar differential rotation and the magnetic activity cycle of the Sun was found. The main question about solar rotation is still not fully resolved, namely the question of what is the physical cause of the solar differential rotation. However, several models have been developed which are more or less supported with observational data. As will be shown in the book, we believe that we can offer a plausible explanation of the solar differential rotation phenomenon, supported with sufficient observational evidence.

The solar differential rotation and its connection with the solar activity cycle and magnetohydrodynamical (MHD) dynamo belong to the few problems in solar physics which are not explained and understood at a satisfactory level yet. To the other complicated problems belong, e.g., solar flares and the problem of the coronal heating. We should, however, not be discouraged. Let us remember that one of the big problems in solar physics, the solar neutrino problem, was successfully solved, at least in principle, within the last few decades.

To illustrate the present-day situation in solar physics we can recall the following picture. Suppose we observe the wood from a distance. Then we zoom in our view, either by using a telescope or by approaching the wood. Suddenly we discover that the wood consists of individual trees, the trees have branches on which there are leaves. Each leaf has its own structure, etc. One next step in the exploration of the wood would be to use the microscope and to analyze the cell structure of the leaf. Another research direction would be to follow how long a tree lives, how it grows, what is the difference between various types of trees, etc. The main point here is that any new discovery expands our knowledge, but also opens new questions, which require additional measurements and analyses. However, we should not be disappointed. At every research step, we should carefully check a new observational result and do our best to explain, understand, and interpret the data and results based on measurements. Of course, the background theory should be in accordance with the generally accepted physics.

The book is structured as follows. In the first introductory chapter, we describe the fundamental properties of the Sun as a star and the procedures how to determine them. At the end, we briefly introduce the topic of the solar rotation. Then follow chapters about the solar rotation elements, coordinate systems, and coordinate transformation. This information is needed for the measurement of the solar rotation and in the next chapter various methods with examples are described in detail. The next topic is the analysis and reduction of measurements, including synodic—sidereal transformation, height correction of observational data, mathematical representation of the solar rotation, statistical data analysis, and error estimates with data filtering. A summary of observational results embraces examples of several extensive studies and their comparisons. In the final chapters of the book, we try to answer the question if we can explain the solar differential rotation. Several theoretical concepts are described and their predictions are compared with reliable observational results. Finally, the topic of the solar rotation and magnetic activity will be addressed and theoretical concepts will be compared with existing observational results, such as temporal variation and north–south asymmetry of the solar rotation. Possible explanations of the experimental results will be presented and discussed.

Throughout the text many references to research papers and books are given, not only to give a proper credit to the authors of the original research, but also to give the hints for further reading. Also, at the end of each chapter exercises and questions are listed. This enables to put more material in the book, besides the main text, and gives an opportunity for the reader to check the understanding of the chapter's content and to perform some additional calculations.

Zagreb, Croatia
Graz, Austria
April 2024

Roman Brajša
Arnold Hanslmeier

Acknowledgements

Roman Brajša would like to thank his teachers who introduced him to solar physics and in particular to the topic of solar rotation investigation: Vladimir Ruždjak (deceased), Bojan Vršnak, and Hubertus Wöhl. They were also supervisors or co-supervisors of my Diploma work and my M.Sc. and Ph.D. Theses, which dealt, completely or partly, with the topic of solar rotation and related phenomena. During my several research stays at the Kiepenheuer-Institut für Sonnenphysik, Freiburg, Germany, I had the unique privilege to discuss several aspects of the solar rotation with some leading experts in the field, such as H. Balthasar, E.-H. Schröter (deceased), and M. Stix. I would also like to thank my colleagues, students, and collaborators, with whom I had the excellent opportunity to work together on interesting topics of solar rotation. The list, which is in no way exhausted, includes in alphabetical order: A. Benz, H. Božić, F. Clette, J. Engler, S. F. Gissot, J.-F. Hochedez, D. Hržina, M. Husak, S. Jejčič, S. Jurač, R. Jurdana-Šepić, A. Kučera, M. Mulec, S. Pohjolainen, I. Poljančić Beljan, J. Rodmann, I. Romštajn, D. Roša, M. Roth, D. Ruždjak, J. Rybák, S. H. Saar, K. Schawinski-Guiton, T. J. Schuck, I. Skokić, D. Sudar, L. Svalgaard, D. Špoljarić, M. Temmer, S. Urpo, G. Verbanac, A. M. Veronig, A. Wegner, and A. Zaatri. Further, I would also like to express my sincere thanks to my late father Pavao Brajša, who wrote more than 35 books about human behavior and human relations, for steady encouragement for my work in science and for writing. From him I have learned how difficult is to write a book, but also how precious and amazing is the delight and joy when the book has been finally published. Finally, I would like to thank my wife Karmen, our daughter Romana and her husband Sven, and my mother Magdalena for their love, patience, and support. Last but not least, I would like to thank our dogs Berna and Belina who patiently accompanied me during working hours and regularly took me out for walks. This enabled me to get some rest between writing and to sharpen my curious mind.

Arnold Hanslmeier would like to thank his colleagues at the Institute of Physics in Graz, namely M. Temmer and A. Veronig and many of the colleagues mentioned already with whom he also has collaborated or is in collaboration. He acknowledges the warm welcomes at the Kiepenheuer Institute in Freiburg and the cooperation with E.-H. Schröter, H. Wöhl, W. Mattig, A. Nesis, P. Brandt, and O. v. d. Lühe. He thanks M. Roth and C. Denker for their cooperation, as well as A. Kučera, J. Rybak, P. Gömöry, and M. Sobotka. He also had research stays at the Observatorie Pic du Midi and especially thanks R. Muller for his hospitality and cooperation. He is also grateful for several common projects and observations with M. Vazquez and J. A. Bonet from the Instituto de Astrofisica de Canarias.

The authors would like to thank the observatories belonging to the institutions of the authors and their staff for producing excellent observational data which are used in our research. These are Hvar Observatory of the Faculty of Geodesy, University of Zagreb, Croatia, and Kanzelhöhe Observatory for Solar and Environmental Research of the Institute of Physics, University of Graz, Austria. In this book, we used many images taken at these observatories and we would like to thank the astronomers at these observatories, especially N. Novak, T. Visković, and J. Čalogović (Hvar Observatory) and A. Schroll, W. Otruba (deceased), and W. Pötzi (Kanzelhöhe Observatory). The authors are very grateful to D. Roša who read the text and made many useful comments.

The authors would also like to acknowledge the financial support from the Austrian–Croatian Bilateral Scientific Project "Multi-Wavelength Analysis of Solar Rotation Profile" (2022–2023), which enabled them several mutual visits while writing this book.

Zagreb, Croatia
Graz, Austria
April 2024

Roman Brajša
Arnold Hanslmeier

Contents

Chapter 1
The Sun as a Star: Fundamental Properties

Fundamental astrophysical properties of a star are the mass, the luminosity, the surface temperature, the radius, and the rotation velocity. In addition, it is important to know the distance to the star, which is in our case the Sun–Earth distance, which is used in the analysis of almost all solar observations. We first introduce these quantities and explain how they can be measured/calculated for the Sun. We also derive the mass–luminosity relation, which relates two of the mentioned fundamental astrophysical quantities of the Sun (or a star). We stress that in this chapter the focus is not on a precise determination of the fundamental astrophysical quantities of the Sun, but rather on the method of measurement/calculation and the line of reasoning. Finally, we introduce the topic of the solar differential rotation and briefly describe its main properties and open questions, which will be addressed later in the book in individual chapters.

1.1 Distance From Earth

We begin with the measurement of the Sun–Earth distance. The reader may want to improve the calculation and to check the values of this and other fundamental quantities addressed in this chapter in the contemporary literature, such as Cox (2000), Stix (2002), and Foukal (2013).

1.1.1 Method: The Third Kepler Law

The distance of the Sun from Earth can be determined using the third Kepler law of planetary motions. This law relates the distances and orbital periods of planets in the solar system. The third Kepler law can be written in the following form:

R. Brajša and A. Hanslmeier, *Solar Rotation*, UNITEXT for Physics,
https://doi.org/10.1007/978-981-97-6879-0_1

$$\boxed{\left(\frac{d_1}{d_2}\right)^3 \approx \left(\frac{P_1}{P_2}\right)^2} \tag{1.1}$$

where d_1 and d_2 are the distances of two planets from the Sun (in fact the lengths of the major semi-axis of the orbital ellipses) and P_1 and P_2 are the orbital periods of these two planets around the Sun. However, Eq. (1.1) solely is not sufficient for the determination of the Sun–Earth distance, since it gives only the ratio of the given quantities and not their absolute values. So, we need to know at least one additional distance within the solar system, to be used as a gauge. This can be performed by measuring the time needed for the radar signal to travel from the Earth to Venus and back in the time of the year when Venus is approximately at the line connecting Earth and the Sun. The radar signal, as any other electromagnetic wave, travels with the speed of light, so that measuring the time directly gives the distance.

We can also raise the question of why we do not measure the Sun–Earth distance by sending and receiving the radar signal directly to the Sun. The reason is in the fact that the Sun does not reflect radar signals well enough at definite heights and moreover it itself emits radio waves making a distinction between the solar radio waves and the reflected signal difficult.

1.1.2 Measurements and Calculation of the Sun–Earth Distance

We now calculate the Sun–Earth distance in km (d_2). Known quantities are $d_2 - d_1 = 41\,400\,000$ km (this is the distance between the Earth and the Venus when both planets are on the same line with the Sun), the orbital period of the Earth around the Sun $P_2 = 365$ days, and orbital period of the Venus around the Sun $P_1 = 225$ days. Solving the equations we get $d_2 = 150\ 000\ 000$ km. This result can be checked by recalling that light, which travels with the speed of about 300 000 km/s, needs 8.3 minutes for traveling from the Sun to the Earth. We note that the Sun–Earth distance is also called the *astronomical unit* A. Today's most precise value of this quantity is $d_2 = A = 149597870 \pm 2$ km.

Finally, we note that there are also some other possibilities to measure distances in the solar system. Here we mention the transit of Venus in front of the solar disk, which enables determination of the Earth–Venus distance if the Venus transit is measured from two different places on the Earth and if the size of the Earth is known. The principle is shown in Fig. 1.1 and we can see that the precise measurements of the times of contacts are crucial for the correct application of the method. Two examples of observation of the Venus transit in the twenty-first century are given in Figs. 1.2 and 1.3. The second possibility is to measure the parallax of an asteroid, e.g., of Eros, from two different positions on the Earth. In this way, the true distance of an asteroid can be determined. An earlier application of the parallax method was the measurement of the Earth–Mars distance in the late seventeenth century. In both

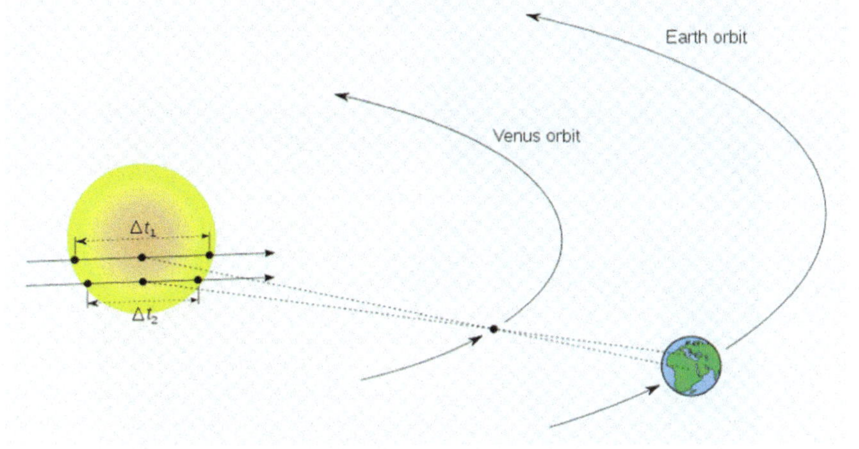

Fig. 1.1 Explanation of the measurement of the Sun–Earth distance using the transit of Venus across the solar disk. Credit: Wikimedia

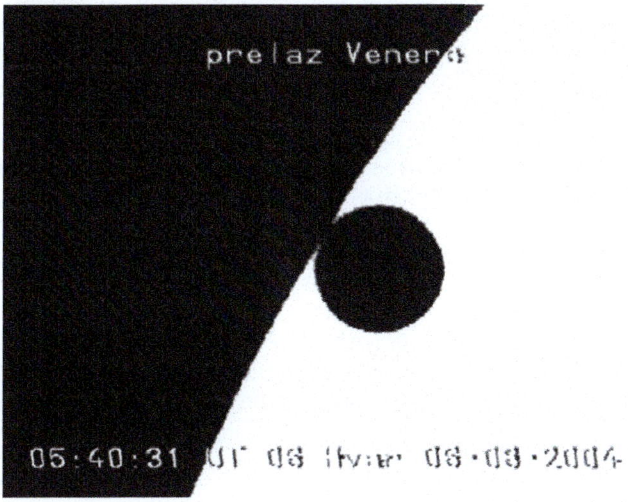

Fig. 1.2 Venus transit across the solar disk on June 6, 2004 recorded with the video system connected to the photospheric solar telescope at Hvar Observatory. The first interior contact is seen. Observation was performed by R. Brajša, N. Novak, and V. Ruždjak. Credit: Hvar Observatory, Faculty of Geodesy, University of Zagreb, Croatia

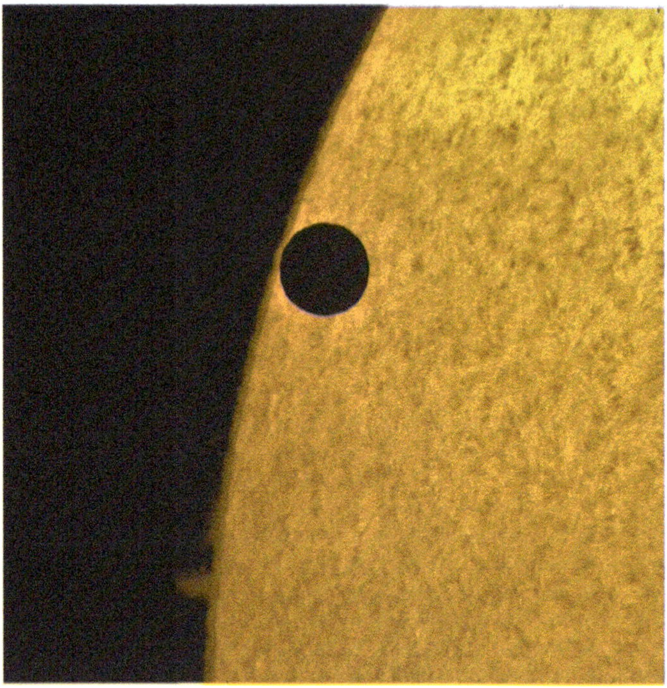

Fig. 1.3 Venus transit across the solar disk on June 6, 2012 recorded with the CCD camera connected to the chromospheric solar telescope at Hvar Observatory. The second interior contact is seen. Observation was performed by J. Čalogović and N. Novak, and the image processing by I. Skokić. Credit: Hvar Observatory, Faculty of Geodesy, University of Zagreb, Croatia

cases, the further procedure is as described earlier in this paragraph. Descriptions of historical measurements of the Sun–Earth distance using the transit and parallax methods can be found in Berry (1961), Hoyle (1962), and Spiering (2021).

1.2 Mass

Unlike the Sun–Earth distance, the solar mass is the first in a series of intrinsic astrophysical quantities of the Sun. It is related to the solar evolutionary status, to its origin, and future development.

1.2.1 Method: The Third Kepler Law

For determination of the solar mass we should use another form of the third Kepler law, which was derived after I. Newton discovered the universal law of gravitation and universal laws of motions of bodies. The universal law of gravitation says that all bodies in the universe are mutually attracted with a force proportional to their masses and inversely proportional to the square of their distance. The first Newton law of motion states that a body remains at rest, i.e., not moving, or that it is uniformly moving on a straight line as long as forces do not act on the body changing any of the mentioned states. The second Newton law states that any time variation of the linear momentum (defined as the product of the body's mass and velocity) is proportional to the force and is acting along the line of force. The third Newton law of motion says that any action is always equal to the force of the same magnitude but in the opposite direction. This could also be formulated simply as action is always equal to the reaction.

Using Newton's universal law of gravitation and second and third Newton's laws of motions the third Kepler law can be written in the following form:

$$\boxed{\frac{d^3}{P^2} = \frac{G}{4\pi^2}(M+m)} \tag{1.2}$$

where d and P have the same meaning as in Eq. (1.1), G is the gravitational constant which amounts to $G = 6.67 \cdot 10^{-11}\,\mathrm{N\,m^2\,kg^{-2}}$, M is the solar mass, and m the mass of a planet to which refer the quantities on the left-hand side of the Equation. We note that the gravitational constant G can be measured with the Cavendish laboratory experiment. We also note that the mass m of any planet in the solar system can be neglected, since it is much smaller than the solar mass.

1.2.2 Calculation of the Solar Mass

Finally, we now calculate the solar mass M using for example astronomical data of the Earth's orbit around the Sun: d_2 given in Sect. 1.1 and the one year's period $P = 365.25$ days. The most precise current value of the solar mass is $M = 1.99 \cdot 10^{30}$ kg. These results can be proven knowing that the mass of the Sun is about 332 000 times larger than the mass of the Earth amounting to $6 \cdot 10^{24}$ kg.

1.3 Luminosity

The solar luminosity is the next intrinsic property of the Sun. It can be determined by measuring the solar constant, for which we will see that it is not really a constant.

1.3.1 The Solar Constant

The total solar luminosity L is in fact the total energy of the Sun radiated per time and it has the physical meaning of the power and is expressed in watts (W). It can be calculated using measurements of the solar irradiance, i.e., the illumination of the Earth with the solar radiation. This physical quantity is also called "the solar constant" S which represents the power of the solar radiation L per area unit at the mean Sun–Earth distance d and is expressed in W/m^2:

$$\boxed{S = \frac{L}{4\pi d^2}.} \tag{1.3}$$

The solar constant can be approximately determined with the simple experiment usually done within the school exercises in astronomy described in, e.g., Kleczek (1987). Here we should bear in mind that approximately one-third of the solar radiation is reflected from the Earth and is going back into the interplanetary space, one-third contributes to the warming of the Earth's atmosphere, and only the final one-third reaches the ground. Moreover, the Earth's atmosphere is blocking many wavelength ranges from the whole solar electromagnetic spectrum. So the best way to determine the solar constant is to measure it directly using devices put on satellites outside the Earth's atmosphere. Using such a device, which captures and measures the energy of almost all photons at various wavelengths coming from the Sun, the solar constant was determined to be $S = 1367 \pm 2\,\text{W}/\text{m}^2$. Finally, we calculate the total solar luminosity using Eq. (1.3) and using the above given values for S and d: $L = 3.8 \cdot 10^{26}$ W.

1.3.2 The Solar Constant is Not Constant

We note that the "solar constant" is not really a constant. Using devices on the satellites small variations of the solar constant were measured during the last four decades. These variations of the solar constant at the Earth's distance amount to about 0.1 % and it is more appropriate to use the term the solar irradiance. It is noteworthy that these variations are directly related to the solar activity cycle, resulting in a slightly larger solar irradiance during the maxima of activity.

1.4 Mass–Luminosity Relation

The two fundamental solar quantities, the mass and the luminosity treated in previous two subsections, are not independent. They are connected through the mass–luminosity relation. Before deriving the mass–luminosity relation the famous Hertzsprung–Russell diagram of stellar properties will be introduced. This diagram relates the stellar luminosity and the surface temperature of the star revealing that any combinations of stellar luminosities and surface temperatures are not possible.

1.4.1 The Hertzsprung–Russell Diagram

The solar quantities introduced in the previous two subsections are mutually related with the famous mass–luminosity relation, representing an empirical relationship which holds for stars on the main sequence in the Hertzsprung–Russell (H-R) diagram. First we say a few words about the Hertzsprung–Russell diagram, one of the most important diagrams in astrophysics. The Hertzsprung–Russell diagram, developed by E. Hertzsprung and H. N. Russell at the beginning of the twentieth century, displays fundamental stellar properties: the spectral class, related to the color index, or logarithm of the temperature on the x-axis and the absolute magnitude, proportional to the logarithm of the luminosity on the y-axis (Fig. 1.4). Please note that the temperature on the x-axis decreases from left to right. The most notable feature of the H-R diagram is its emptiness; not all possible combinations of stellar temperatures and luminosities are observed. The great majority of stars are placed in a couple of good defined regions in the diagram. First we have the main sequence extending from the upper-left to the lower-right parts of the diagram. This correlation means that for the stars on the main sequence increasing the temperature represents also increasing the luminosity. Other important regions in the diagram are populated with red giants, supergiants, and white dwarfs. It is also noteworthy that the range of stellar luminosities is much larger than the range of the stellar masses.

The mass–luminosity relation states

$$\boxed{L \propto M^{3.5}\ .} \tag{1.4}$$

1.4.2 Derivation of the Mass–luminosity Relation

Now we try to physically understand this empirical relation and the easiest way to derive it is by using the dimensional analysis. We follow the derivation according to Bahcall (1989). We start with the equation for energy transport:

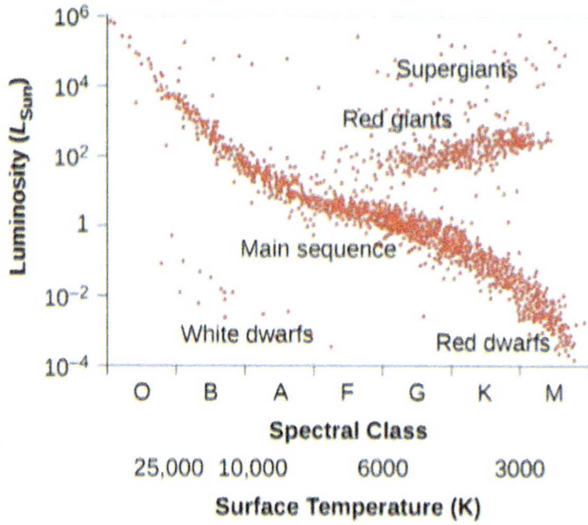

Fig. 1.4 The Hertzsprung–Russell diagram. Adapted from Lumen Astronomy

$$L_r \propto r^2 \frac{1}{\kappa\rho} \frac{\mathrm{d}T^4}{\mathrm{d}r} \tag{1.5}$$

where L_r is the energy per unit time which passes through a sphere of radius r and T is the temperature. The total opacity κ is the sum of a radiative and conductive contributions. From this equation we get the following proportionalities:

$$L_r \propto \frac{R^2T^4}{\kappa\rho R} \propto \frac{R^4T^4}{\kappa M} \, . \tag{1.6}$$

We note that r, which denotes any radial distance in a star, is here replaced by R denoting the stellar radius. Now we recall the equation of state, i.e., the equation of hydrostatic equilibrium:

$$\frac{\mathrm{d}P(r)}{dr} = -\frac{GM(r)\rho(r)}{r^2}. \tag{1.7}$$

From this equation of state we get the following relation:

$$T \propto \frac{P}{\rho} \propto \frac{M\rho}{R\rho} \propto \frac{M}{R} \, . \tag{1.8}$$

Combining Eqs. (1.6) and (1.8) we get the mass–luminosity relation in its most simple, crude form:

$$L \propto M^3 \tag{1.9}$$

with a slightly different exponent from the empirical relation (1.4), but still rather close. The conclusion from this simple derivation is that we can understand one of the fundamental stellar properties from the basic physical principles. We note that the mass–luminosity relation is valid only for the main sequence stars.

1.5 Surface Temperature

The surface temperature of the Sun, which we already met in the context of the Hertzsprung–Russel diagram, also belongs to the intrinsic solar quantities. For its determination Planck's law of radiation should be introduced first.

1.5.1 Planck's Radiation Law

One of the possible procedures for the determination of the surface temperature of the Sun is based on Planck's radiation law. This fundamental physical law from quantum mechanics mathematically describes the distribution of the emitted energy of a black body in time unit and according to the radiation wavelength. Planck's law can be written as

$$\boxed{I_\lambda(T) = \frac{2hc^2}{\lambda^5}\frac{1}{\exp(hc/k\lambda T) - 1},} \tag{1.10}$$

where I_λ is the intensity of radiation at the given wavelength λ and the temperature T. Further, h is the Planck constant ($h = 6.626 \cdot 10^{-34}$Js), c the speed of light ($c = 2.9979 \cdot 10^8$m/s), and k the Boltzmann constant ($k = 1.3806 \cdot 10^{-23}$J/K). A black body is the physical body which emits and absorbs radiation, but which is in the state of the thermal equilibrium with its surroundings. So, neither the received nor the emitted radiation influences this equilibrium. For various black body temperatures, the distribution of the radiation intensity according to the wavelength can be calculated. This quantity $I(\lambda)$ is usually represented by Planck curves: the wavelength is given at the x axis and the intensity of radiation in unit time on the y axis (Fig. 1.5). It is important to note that the Planck curve of the black body with a higher temperature has its radiation maximum at shorter wavelengths than the Planck curve of the black body with a lower temperature. Moreover, the total radiation power, integrated over all wavelengths, is larger, which is described by Wien's law:

$$T\lambda_{\text{max}} = 2.98 \times 10^{-3}\,\text{m}. \tag{1.11}$$

Wien's law can be derived from Planck's law.

1.5.2 Determination of the Solar Surface Temperature

Using the above-mentioned properties and knowing that the maximum of the solar radiation falls into the yellow part of the spectrum, at the transition to the green part, we can determine the Sun's surface temperature T. It is important to note that Planck's curves (Fig. 1.5) do not intersect, so that the relationship maximal intensity vs. wavelength is established in a unique way, allowing an unambiguous determination of the temperature. For the Sun we get $T = 5780$ K which is in fact called the effective solar temperature, since the Sun is not a perfect black body. This value is close to the temperature at the solar surface, i.e., in the photosphere. The term "surface" should be used with caution, since the Sun has no solid surface, as for instance the Earth. The Sun is composed of plasma, i.e., of ionized gas. We also note that above the photosphere the temperature first slightly drops to the so-called temperature minimum (a few hundred K less, at a few hundred km above the photosphere), then rises to some 10,000 K in the chromosphere (the next layer in the solar atmosphere), and finally steeply grows up to 1–2 million K in the uppermost layer called the corona.

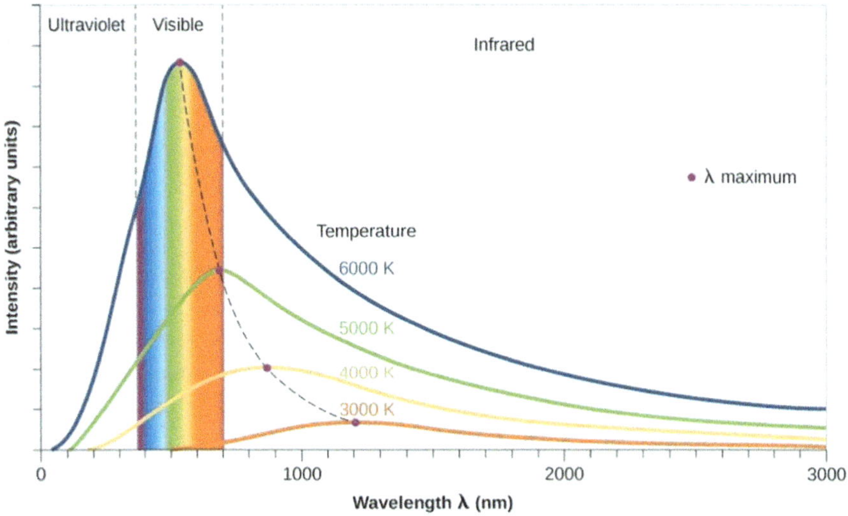

Fig. 1.5 The energy distribution of radiation as a function of the temperature of the object according to Planck's radiation law. Adapted from Lumen Astronomy

1.6 Radius

The solar radius is another astrophysical quantity of the Sun defining its property as a star. It can be measured knowing the Sun–Earth distance, introduced at the beginning of this chapter.

1.6.1 The Geometrical Method

The solar radius R can be calculated from its apparent radius on the sky given in arc units (degrees, minutes, seconds) α and the Sun–Earth distance d:

$$R = d \times \tan \alpha. \tag{1.12}$$

We note that it is extremely dangerous to look at the Sun with the naked eye, and especially through the telescope, because the strong, concentrated Sun's light can seriously damage human's eye and vision/eyesight. Astronomers overcome this hurdle generally in two ways. The first method is to observe the Sun in projection, on the white paper placed on a screen located behind the telescope's ocular (see Fig. 1.6).

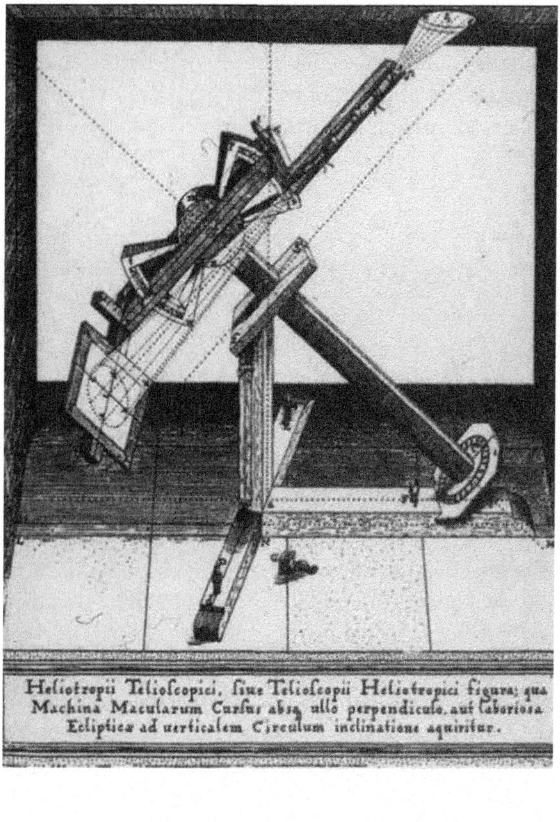

Fig. 1.6 Projection method for solar observation as used by Christoph Scheiner

The second method is the usage of various filters with the telescope, which significantly diminish solar radiation. Using the projection method the apparent diameter (and consequently the radius) of the Sun can be measured in, e.g., cm and subsequently in angular units using simple laws of geometrical optics. Then it is easy to calculate the true solar radius using Eq. (1.12).

1.6.2 Calculation of the Solar Radius

The current most precise value is $R = 6.96 \cdot 10^8$ m. We can check the value of the solar radius with the following simple exercise. It is well known that the apparent diameters of the Sun and the Moon, as seen from the Earth, are approximately the same. With this observational fact the total solar eclipse is explained: the Moon shifts in front of the Sun and completely occults the solar disk. Angular diameter of the Moon on the sky is about half a degree which is very similar to the angular diameter of the Sun. Using $\alpha = 15'$ and the known Sun–Earth distance d, the solar radius can easily be calculated with Eq. (1.12). The obtained result can be checked knowing that the true solar radius is about 109 times larger than the radius of the Earth which is equal to about 6400 km.

Modern precise measurements of the solar radius are very complicated, because the Sun is very bright and does not have a sharp edge. However, these measurements, including temporal variations of solar radius and oblateness of the Sun, are rather important in contemporary astrophysics, because of their possible relation to the solar activity cycle and irradiance variations (Rozelot (2003) and Rozelot and Neiner (2009)).

1.7 Solar Rotation: An Introduction

Solar rotation completes the list of fundamental properties of the Sun as star. In this section we briefly introduce the main topics of this book (Fig. 1.7).

1.7.1 Solar Rotation as a Fundamental Property of the Sun

The final fundamental property of the Sun as a star is the *solar rotation*. However, this physical quantity cannot be represented by a single number, as was the case with the properties described in previous subsections. The rotational speed at the solar equator is about 2 km/s. For comparison we note that the rotational speed of the Earth at the equator amounts to 465 m/s. With increasing solar latitude this speed is decreasing, not only because of the property of any rotating sphere, but also since the angular velocity is decreasing with increasing latitude, which is possible only in the gaseous

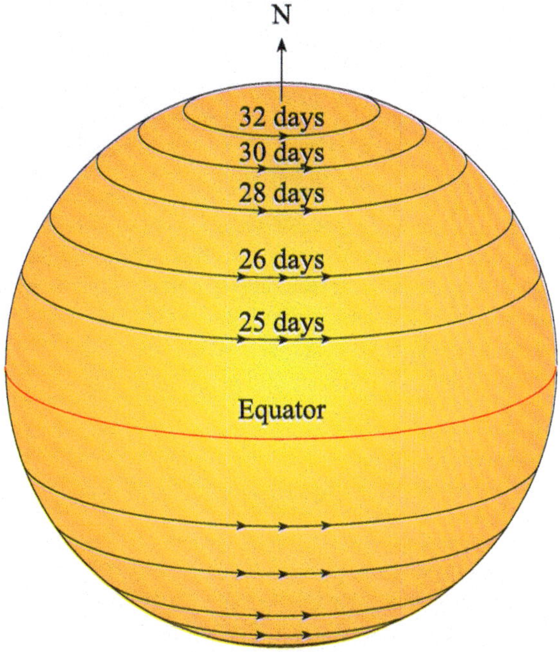

Fig. 1.7 Illustration of the solar differential rotation. Credit: D. Hržina

bodies. On the other hand, solid bodies, like the Earth, have the same angular velocity at all latitudes (this is in principle an approximation, but valid to the very high level). Let us consider two cities lying on the same longitude on the Earth; they remain on the same longitude over days, months, years, etc. The Sun, however, is composed of plasma and fluid motions in the rotating sphere allow differential rotation, i.e., the latitudinal dependence of the angular rotational velocity (Fig. 1.10). A general introduction to the topic of the solar differential rotation can be found in following publications: Howard (1984), Schroeter (1985), Snodgrass (1992), Howard (1996), Beck (2000), and Stix (2002).

1.7.2 Synodic and Sidereal Solar Rotation

Now we introduce the concept of *synodic and sidereal solar rotation.* The synodic solar rotation is measured from the Earth and gives the rotational motion of the Sun compared to the Earth which is moving around the Sun in the same direction as the Sun itself rotates. For this reason, this is the apparent solar rotation which is not the same as the true rotation, which is the solar rotation respectively to the stars. This other quantity is called the sidereal solar rotation. The angular rotation of the Sun is usually denoted by ω and we subsequently denote the synodic solar rotation by ω_{syn}

and the sidereal solar rotation by ω_{sid}. From obvious reasons $\omega_{sid} > \omega_{syn}$. A simple approximate transformation between the two values is given by

$$\omega_{sid} = \omega_{syn} + \frac{360^0}{365\,\mathrm{d}} \,. \tag{1.13}$$

So, a simple but approximative way of synodic–sidereal transformation is by adding a constant, 0.99 degrees per day, to the synodic rotational velocity to get the sidereal one. This approximation assumes that the Earth has a constant speed around the Sun, which we know is not true, and uses a rounded number of days for the length of the year. Later in the book we will derive the exact equation for this transformation using the second Kepler law. As we will see, the correct formula is season's dependent and the date of observation should be taken into account.

1.7.3 Variability of the Solar Rotation

Generally speaking, the solar rotation is a *function of the latitude, height/depth, and time*. As we have seen, the latitudinal dependence means decreasing of the rotational velocity with increasing latitude. In the first approximation the rotational profile is symmetric around the solar equator and the angular velocity decreases in the same way towards both solar poles. This is, however, not true and significant differences from the north–south symmetry were found, as we will see later in this book. Further, the solar rotation is a function of the height, i.e., the layer in the solar atmosphere where the measurements are performed. Similarly, it is also a function of the depth below the solar surface. Finally, the solar rotation is also a function of time, on various scales. Our main interest will focus on the temporal variations of the solar rotation related to the phase of the solar 11-year cycle. The solar cycle is in fact the cycle of the solar magnetic activity and we can consider these variations of the solar rotation to be connected with the magnetic field. Moreover, secular changes of the solar rotation were also reported, as well as indications of a different behavior of the solar rotation during the overall decreasing of solar activity during the Maunder minimum in the seventeenth and eighteenth centuries.

1.7.4 A Brief History of Solar Rotation Research

We now briefly describe the *history of discovery of solar differential rotation*. Solar rotation and its differential nature (the dependence on the latitude) have been known since the discovery of sunspots. Spots on the Sun, or sunspots, were discovered independently by four astronomers at more or less the same time, at the beginning of the seventeenth century. These astronomers are Fabricius (Johannes Goldschmidt) (1587–1615) in Holland, Galileo Galilei (1564–1642) in Italy, Thomas

Harriot (1560–1621) in England, and Christoph Scheiner (1575–1650) in Germany. Scheiner observed the image of the Sun projected on the screen (with paper) placed behind the telescope tube (Fig. 1.6). In this way, he also avoided direct observation of the Sun through the telescope, which is very dangerous for the eyesight. At the beginning of the seventeenth century Scheiner discovered that sunspots at higher latitudes rotate with a smaller angular velocity than the sunspots near the equator. German astronomer Hermann Vogel (1841–1907) from Potsdam discovered in 1871 a redshift of the Fraunhofer spectral lines at the west solar limb and a blueshift of those lines at the east solar limb. This phenomenon was explained by the Doppler effect, clearly indicating solar rotation. Scheiner and Vogel introduced the two methods for the measurement of the solar rotation, the tracer method and the spectroscopic method, respectively. The methods for the determination of the solar rotation will be described in the next subsection.

1.7.5 Methods for Solar Rotation Measurements

The most important *methods for the measurement of the solar rotation* are the *tracer method*, the *Doppler method*, and *helioseismology*. The tracer method consists of following some discernible objects (tracers) in the solar atmosphere during their motion along with solar rotation. The positions of tracers on the visible solar disk are measured each day and the rotation velocity is calculated by dividing the distance with the elapsed time. An example of how modern devices are used for the tracing objects on solar disk is given in Figs. 1.8 and 1.9. The most natural first choice for the tracer are the sunspots and historically they were used as first tracers at the beginning of the seventeenth century (see the preceding subsection). Later also other objects were considered and used, such as solar prominences, faculae, various chromospheric structures, coronal bright points, coronal holes, magnetic patterns, etc. In addition to the visible part of the solar spectrum, other spectral ranges are also used, such as UV, EUV, and X-ray, as well as infrared, mm, and radio parts. With higher observing cadence, especially with telescopes mounted on the satellites, displacements on the shorter time scales, less than one day, can be precisely measured. In the twentieth century the tracer method was further developed, embracing the following special cases/applications: the measurement of the time between successive passages of identified solar structures across the central meridian of the Sun, and the cross-correlation and autocorrelation methods. These procedures enabled the determination of solar rotation using almost all recognized patterns/objects on the Sun. The measurement of the spectral line shifts of portions of the solar disk moving towards the observer or away from it enables determination of the relative velocity of the Sun respectively to the observer, using the Doppler effect. After carefully taking into account all relative motions (including Earth's rotation and revolution) the solar rotation velocity can be determined.

A further development of the Doppler (also called the spectroscopic) method in the second half of the twentieth century led to the new branch of solar physics, the

Fig. 1.8 Patrol telescopes for solar observations at the Kanzelhöhe Observatory for Solar and Environmental Research (Institute of Physics, University of Graz, Austria)

helioseismology. This method is based on the precise measurements of the spectral line oscillations on the Sun. There are many applications of this method including the determination of the rotation velocity below the solar surface.

The solar differential rotation is most usually *represented* by the following formula:

$$\boxed{\omega(b) = A + B\sin^2 b + C\sin^4 b\ ,} \tag{1.14}$$

where ω is the solar angular rotation velocity (expressed in deg/day), b the heliographic latitude (expressed in deg), and A, B, and C the solar rotation parameters. The solar rotation parameter A represents the rotation velocity at the solar equator, while the parameters B and C represent the gradient of the solar differential rotation, i.e., its dependence on the latitude. The parameter C is important for high solar latitudes and can be neglected for low latitude tracers, such as sunspots. The angular

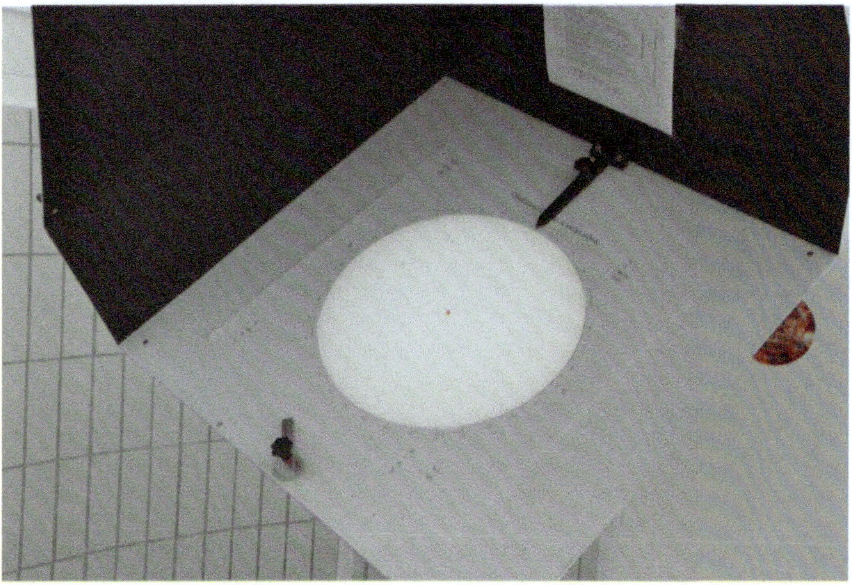

Fig. 1.9 Projection device for making the draft of sunspot drawings. The enlargement of the solar disk to a diameter of about 25 cm corresponds to a magnification factor of about 100. Adapted from Poljančić Beljan et al. (2017)

solar rotation velocity is expressed in deg/day, in μrad/s, or in nHz. It can also easily be transformed into rotational period in days.

For sunspots, which rarely appear at latitudes larger than 40 deg, both in northern and southern solar hemispheres, the last term in Eq. (1.14) can be neglected and we set $C = 0$. One example of the solar differential rotation profile determined by sunspot observations is given in Fig. 1.10 and the sidereal parameters which define the differential rotation curve are $A = 14.52 \pm 0.01$ deg/day and $B = -2.80 \pm 0.09$ deg/day (Poljančić Beljan et al. (2017) and Ruzdjak et al. (2018)).

1.7.6 Can We Explain the Solar Differential Rotation?

We now briefly address the question of if we can *explain and understand* why the Sun rotates differentially. This is a rather difficult question and the whole answer is not known yet. It is obvious that the differential rotation is possible in a rotating sphere composed of a fluid–plasma in the solar case. Still, the following question remains open: what is the physical cause of the differential rotation and more specifically, why the profile of the solar differential rotation is as it is. As we have seen, on the Sun the angular rotation velocity decreases from equator to the poles, on both sides of the equator. Any effort to explain solar differential rotation should take into account

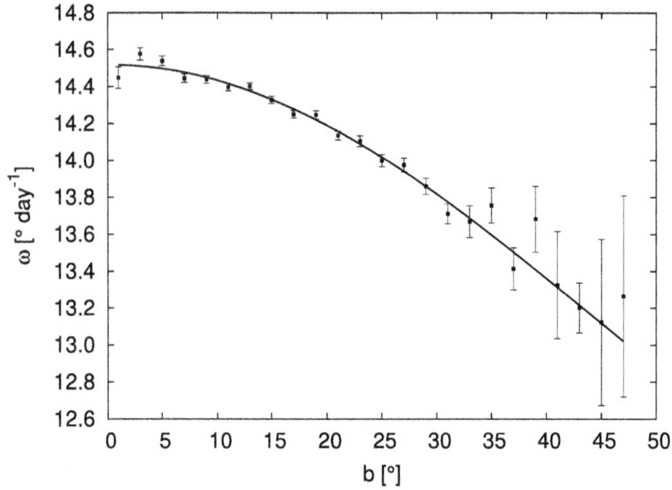

Fig. 1.10 Differential rotation profile obtained from tracing sunspot in the period 1964–2016. Data were recorded at the Kanzelhöhe Solar Observatory. Points with error bars are averaged 2 deg latitude bin values and the best-fit profile is shown with the solid line. Adapted from Ruzdjak et al. (2018)

at least two important physical properties: conservation of angular momentum and Coriolis force. Moreover, besides pure hydrodynamics, magnetic field cannot be neglected and the analysis should include the full palette of magnetohydrodynamical (MHD) effects. To maintain the observed profile of the solar differential rotation angular momentum should be transferred from higher latitudes towards the equator. The main possible candidates as angular momentum transporters are meridional motions, correlations of velocity components (Reynolds stresses), and giant cell circulations. Later in the book we will compare the results of theoretical analysis and modeling with observational results. In this way we can discriminate between various models and try to answer the question from the beginning of this paragraph.

1.7.7 Why it is Important to Study Solar Differential Rotation?

Finally, we can ask the question of *why it is important* to study solar differential rotation. Measurements of the solar differential rotation provide an observational constraint on theoretical models of the rotating convection zone and the solar MHD dynamo, which plays an important role in generating and maintaining the solar magnetic field. This magnetic field is responsible for appearance of almost all activity phenomena on the Sun and its cyclic variability has a period of about 11 years. We speak about the well-known 11-year solar activity cycle. According to present conceptions, solar differential rotation enables and drives the 11-year cycle of solar magnetic activity. Understanding the solar activity cycle, which is driven by a MHD dynamo and differential rotation, remains a key unsolved problem in solar physics, along with, e.g., solar flares and the heating of the solar corona. There is also a practical aspect of the solar rotation research, related to the space weather. Plasma flows from the Sun into the interplanetary space from the sources in the solar atmosphere which rotate with the Sun. These flows are directional and an obvious question is if the plasma cloud will hit a planet or a spacecraft in the solar system. Knowing the rotational speed of various solar structures may help in modeling space weather conditions and their variability at places of interest in the solar system.

Exercises and Questions

1. Calculate the Sun–Earth distance using the data for Mars: the Earth–Mars distance when Mars is in opposition and its orbital period around the Sun. The Earth–Mars distance, when Mars is in opposition, varies because the orbits of both planets are elliptical. For the calculation use the average value of 80 million km.

2. What is the fraction of combined masses of all 8 planets in the solar system compared with the mass of the Sun?

3. Calculate the "solar constant" at the surface of Mars.

4. At what place in the H-R diagram is the Sun?

5. Derive the equation of state Eq. (1.7) taking into account the condition for the hydrostatic equilibrium which states that the outward net force from the gas pressure must be balanced by the inward force given by the gravitational acceleration due to the weight of a mass element.

6. What color have the hottest stars?

7. A total solar eclipse is possible since the Sun is further away from the Earth, as compared to the Moon, by the same factor by which the true solar diameter is larger than the true diameter of the Moon. What value has this factor?

8. What is the sidereal angular rotation velocity at the solar equator measured by sunspots? Transform this value to the synodic one. What is the sidereal angular rotation velocity determined by sunspots at latitudes $b = 20$ deg and $b = 40$ deg? Use data from text and check the results with the solar differential rotation profile given in Fig. 1.10.

References

Bahcall, J.N.: Neutrino astrophysics. Cambridge University Press (1989)

Beck, J.G.: A comparison of differential rotation measurements-(Invited review). Solar Phys. **191**(1), 47–70 (2000)

Berry, A.: A short history of astronomy, from earliest times through the nineteenth century. Dover Publications Inc, New York (1961)

Cox, A.N. (ed.): Allen's astrophysical quantities, 4th edn. AIP Press, Springer, New York (2000)

Foukal, P.: Solar astrophysics, 3rd revised edn. Wiley-VCH-Verlag (2013)

Howard, R.: Solar rotation. Ann. Rev. Astron. Astrophys. **22**, 131–155 (1984)

Howard, R.F.: Solar active regions as diagnostics of subsurface conditions. Ann. Rev. Astron. Astrophys. **34**, 75–110 (1996)

Hoyle, F.: Astronomy. MacDonald, London (1962)

Kleczek, J. (ed.), Minnaert, M.G.J.: Exercises in astronomy. D. Reidel Publ, Comp (1987)

Poljančić Beljan, I., Jurdana-Šepić, R., Brajša, R., Sudar, D., Ruždjak, D., Hržina, D., Pötzi, W., Hanslmeier, A., Veronig, A., Skokić, I., Wöhl, H.: Solar differential rotation in the period 1964–2016 determined by the Kanzelhöhe data set. Astron. Astrophys. **606**, A72 (2017)

Rozelot, J.-P. (ed.).: The sun's surface and subsurface: Investigating shape and irradiance, vol. 599. Springer-Verlag, LNP (2003)

Rozelot, J.-P., Neiner, C. (eds.).: The rotation of sun and stars, vol. 765. Springer-Verlag, LNP (2009)

Ruzdjak, D., Sudar, D., Brajsa, R., Skokic, I., Poljancic Beljan, I., Jurdana-Sepic, R., Hanslmeier, A., Veronig, A., Poetzi, A.: Meridional motions and reynolds stress determined by using kanzelhöhe drawings and white light solar images from 1964 to 2016. Solar Phys. **293**(4), 59 (2018)

Schroeter, E.H.: The solar differential rotation-Present status of observations. Solar Phys. **100**, 141 (1985)

Snodgrass, H.B.: Synoptic observations of large scale velocity patterns on the sun. In: Harvey, K.L. (ed.) The Solar Cycle, vol. 27 of Astronomical Society of the Pacific Conference Series, pp.205 (1992)

Spiering, C.: Neutrino-astronomie. Springer-Verlag, Berlin (2021)

Stix, M.: The sun: An introduction. Springer-Varlag, Berlin (2002)

Chapter 2
Solar Rotation Elements

The starting point in the analysis of the solar rotation is the determination of the solar rotation axis in space, which is described by the solar rotation elements. First, we define the two solar rotation elements, the angle between the ecliptic plane and the equatorial plane of the Sun (i) and the ecliptic longitude of the ascending node of the solar equator (Ω), and then introduce the concept of the Carrington elements of the solar rotation. Then we describe how the solar rotation elements can be determined. The needed coordinate transformation is derived using two procedures: the first one based on the spherical trigonometry and the second one based on linear algebra and analytical geometry. Finally we present the results of the solar rotation elements measurements from the seventeenth century up to the present time.

2.1 The Position of the Solar Rotation Axis in Space

In this section we introduce the solar rotation elements and give their values as determined by R. C. Carrington.

2.1.1 Definition of the Solar Rotation Elements

The position of the solar rotation axis in space is defined with two rotation elements. These are the angle between the ecliptic plane and the equatorial plane of the Sun (i) and the ecliptic longitude of the ascending node of the solar equator (Ω). In other words, the angle Ω is the angle between the spring equinox direction and the direction where the solar equator intersects the ecliptic plane from south to the north. The graphical representation of the meaning of the solar rotation elements is given in Fig. 2.1.

R. Brajša and A. Hanslmeier, *Solar Rotation*, UNITEXT for Physics,
https://doi.org/10.1007/978-981-97-6879-0_2

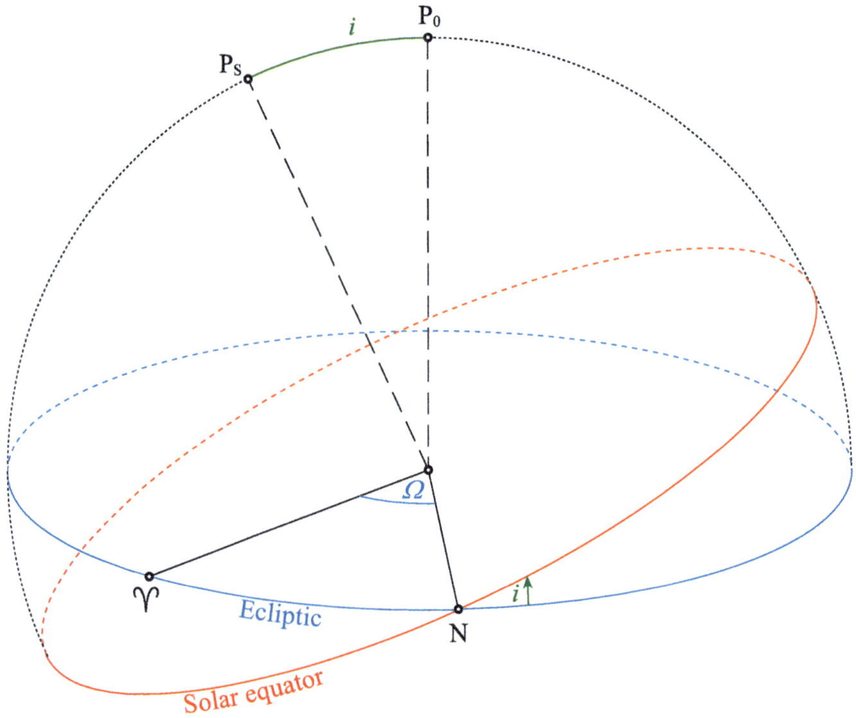

Fig. 2.1 The Carrington elements of the solar rotation i and Ω. Credit: D. Hržina

For easier understanding of this relatively complicated geometrical concept we shall use a simple visualization. Imagine three planes in space: (a) the plane of the celestial equator, which is the projection of the Earth's equator into the space, (b) the ecliptic plane, which is the plane in which the Earth orbits around the Sun, and (c) the plane of the solar equator. The obliquity of the ecliptic ε and the direction of the spring equinox are defined by the intersection of the celestial equator plane and the ecliptic plane. The angle i between the ecliptic plane and the solar equatorial plane is determined by the intersection of these two planes. Knowing the angle i we still do not know in which direction points the solar rotation axis. The answer gives the angle Ω, i.e., the angle between the spring equinox direction, defined earlier, and one of the intersections of the solar rotation plane and the ecliptic plane.

2.1.2 The Carrington Elements of the Solar Rotation

The first estimate of the solar rotation element i was given by Christoph Scheiner (1630) already in 1630. In spite of several measurements thereafter, the first precise values of the both solar rotation elements were measured and derived by R. C. Carrington (1863) in 1863. His values are

$$\boxed{i = 7.25^\circ} \tag{2.1}$$

$$\boxed{\Omega = 73.67^\circ + (Y - 1850) \times 0.01396^\circ} \tag{2.2}$$

where Y stands for the calendar year for which the value of Ω refers to. From Eq. (2.2) we see that the value of Ω increases by 0.01396° per year which is caused by the axial precession of Earth. The quantities i and Ω are called the Carrington elements of the solar rotation and more details about them can be found in Woehl (1978), Stark (1979), Stark and Woehl (1981), Balthasar et al. (1986), Balthasar et al. (1987), and Cox (2000).

2.2 Determination of the Solar Rotation Elements

For the determination of the solar rotation elements the coordinate transformation is needed, which is derived here using two procedures: the first one based on the spherical trigonometry and the second one based on linear algebra and analytical geometry. At the end of this chapter we present the results of the solar rotation elements measurements from the seventeenth century up to the present time.

2.2.1 *Coordinate Transformation Based on Spherical Trigonometry*

The aim of this subsection is to describe how the celestial equatorial coordinates (declination and right ascension) of an object on the solar disk can be determined from the known coordinates of the solar disk center and the measured apparent position of this object on the solar disk. The final aim is to develop a procedure for determination of the solar rotation elements i and Ω from observations. We follow the derivation as described in detail in Roša et al. (2021). To avoid possible misunderstanding we note that the solar coordinates, latitude and longitude, will be treated in the next chapter.

In Fig. 2.2 a drawing of the solar disk with the position of an object on it (e.g., sunspot) is shown. The line WE, passing through the center of the solar disk, is parallel to the plane of the celestial equator. The angular distance of the sunspot from the direction WE corresponds to the difference between the declination of the sunspot δ_{f} and the declination of the center of the solar disk δ_{S}. The right ascension of the sunspot α_{f} is equal to

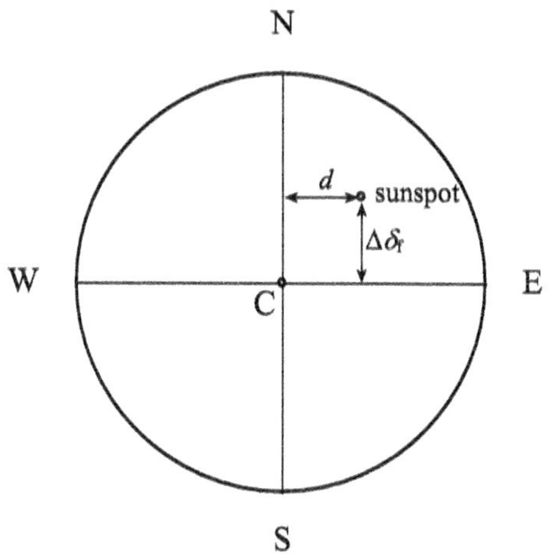

Fig. 2.2 A sketch of the solar disk with the position of an object on it, e.g., a sunspot. The line WE is parallel to the plane of the celestial equator and can be easily determined from observations. Adapted from Roša et al. (2021)

$$\alpha_f = \alpha_S + \frac{d}{\cos \alpha_f} \tag{2.3}$$

where d is the angular distance of the sunspot from the arc NS which is perpendicular to WE and which passes through the center of the solar disk, and α_S is right ascension of the solar disk center.

From the quantities α and δ we can easily calculate topocentric ecliptic coordinates of the sunspot. These are the topocentric ecliptic latitude β_f' and the topocentric ecliptic longitude λ_f'. The well-known formulae for the transformation between the equatorial and ecliptic coordinate systems are used:

$$\sin \beta_f' = \cos \varepsilon \sin \delta_f - \sin \varepsilon \cos \delta_f \sin \alpha_f \tag{2.4}$$

$$\cos \beta_f' \cos \lambda_f' = \cos \delta_f \cos \alpha_f \tag{2.5}$$

$$\cos \beta_f' \cos \lambda_f' = \sin \varepsilon \sin \delta_f + \cos \varepsilon \cos \delta_f \sin \alpha_f \ , \tag{2.6}$$

where ε is the obliquity of the ecliptic, i.e., the angle between the plane of the ecliptic and the celestial equator, $\varepsilon = 23.44$ deg.

We now describe the procedure for the determination of the elements of the solar rotation. With the known obliquity of the ecliptic ε at the observation time t and from Eqs. (2.4)–(2.6) we can calculate the apparent topocentric ecliptic coordinates of the sunspot, λ_f' and β_f'. The apparent angular, λ_S' and β_S', and rectangular, x_S, y_S, and z_S, topocentric coordinates of the Sun as well as the current distance to the Sun can be determined using some of the ephemeris data publicly available.

For further calculation the distances and angles defined in Fig. 2.3 are used. Since. the apparent angular distance of the sunspot from the solar disk center, σ, is small, the following relation can be used:

$$\sigma = 2 \arcsin\sqrt{\sin^2 \frac{\beta_f' - \beta_S'}{2} + \cos\beta_f' \cos\beta_S' \sin^2 \frac{\lambda_f' - \lambda_S'}{2}}. \quad (2.7)$$

According to Fig. 2.3 the heliocentric angular distance between the sunspot and observer is equal to

$$\sigma' = 180° - \sigma - \arcsin\left(\sin\sigma \cdot \frac{d_{\text{observer}-S}}{r_S}\right), \quad (2.8)$$

where $d_{\text{observer}-S}$ is the topocentric distance to the solar disk center and r_S is the radius of the Sun. The topocentric distance to the sunspot, d_{f-S}, is

$$d_{f-S} = r_S \frac{\sin\sigma'}{\sin\sigma}. \quad (2.9)$$

The heliocentric rectangular ecliptic coordinates of the sunspot are

$$x_f = d_{f-S} \cos\lambda_f' \cos\beta_f' - x_S, \quad (2.10)$$

$$y_f = d_{f-S} \sin\lambda_f' \cos\beta_f' - y_S, \quad (2.11)$$

$$z_f = d_{f-S} \sin\beta_f' - z_S. \quad (2.12)$$

Finally we get the heliocentric ecliptic coordinates of the sunspot (note that the subscripts are now dropped out):

$$\beta = \arcsin\frac{z_f}{r_S}, \quad (2.13)$$

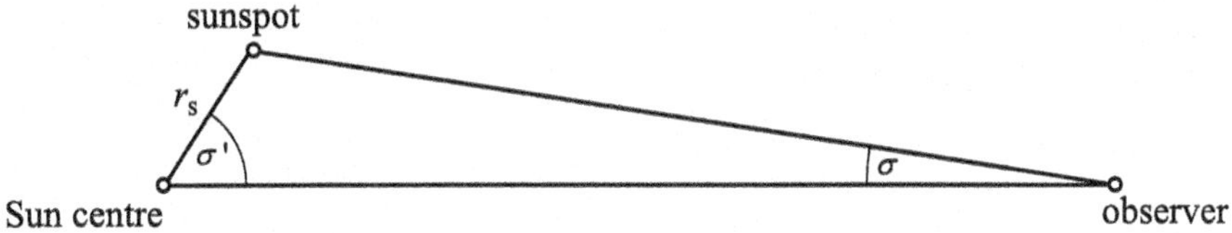

Fig. 2.3 A sketch of the relative positions of the distance between the observer and the Sun, $d_{\text{observer}-S}$, the apparent angular distance between the solar disk center and the sunspot, σ, and the solar radius, r_S. The aim is to determine the angle between the lines $d_{\text{observer}-S}$ and r_S. Adapted from Roša et al. (2021)

$$\lambda = \arcsin \frac{x_f}{y_f} \; . \tag{2.14}$$

The value of the obliquity of the ecliptic is used for the epoch of observation and the same value is used for the coordinate transformation from the equatorial system to the ecliptic one.

The heliocentric celestial sphere is shown in Fig. 2.4. The position of a sunspot is denoted with X, having the coordinates λ, β in the celestial heliocentric ecliptic system and coordinates l, b in the heliographic system (the latitude and the longitude, respectively). The longitude l is the arc NX', where N is the ascending node of the solar equator. The point P_0 is the north ecliptic pole, and the point P_S is the north pole of the solar rotation axis.

The direction of the solar axis in space can be uniquely determined by two elements: the inclination i and the ecliptic longitude of the ascending node of the solar equator Ω (see Fig. 2.1). The angle i is the angle between the ecliptic plane and the solar equatorial plane. The angle Ω is the angle, in the ecliptic plane, between the equinox direction and the direction where the solar equator intersects (N on Fig. 2.2) the ecliptic from the south, i.e., in the sense of rotation.

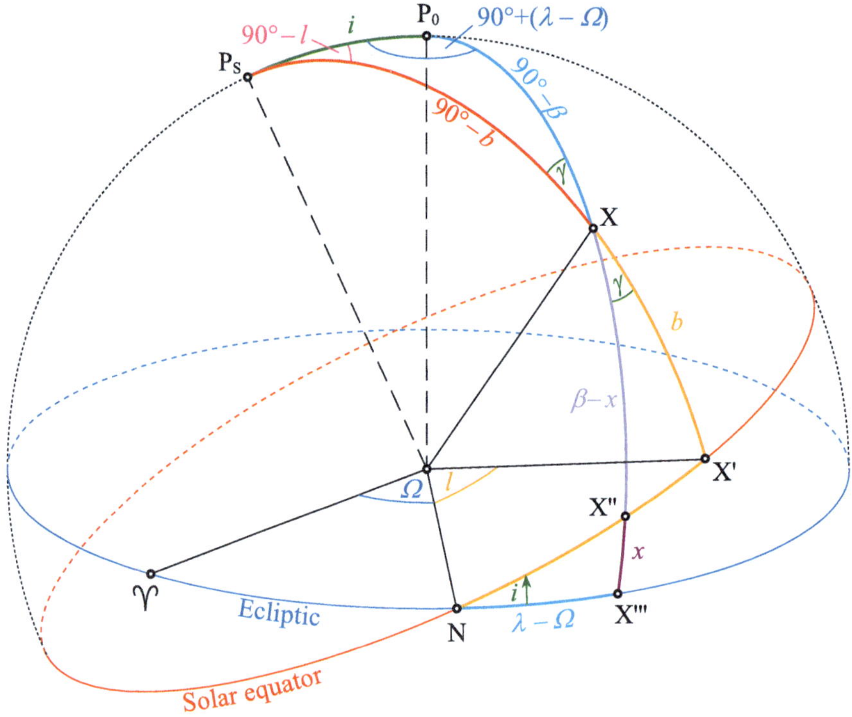

Fig. 2.4 Coordinates of an object on the Sun X (e.g., the sunspot) on the heliocentric celestial sphere, both in the ecliptic and the heliographic coordinate systems. Adapted from Roša et al. (2021)

Now we apply the basic formulae of spherical trigonometry to the spherical triangle $P_S X P_0$ (see Fig. 2.4) to get

$$\cos i = \cos(90^\circ - b)\cos(90^\circ - \beta) + \sin(90^\circ - b)\sin(90^\circ - \beta)\cos\gamma\ , \quad (2.15)$$

from which we find the following expression:

$$\cos\gamma = \frac{\cos i - \sin b \sin\beta}{\cos b \cos\beta}\ . \quad (2.16)$$

From the triangle XX'X" we get

$$\cos\gamma = \tan b \cot(\beta - x) = \tan b \frac{1 + \tan\beta \tan x}{\tan\beta - \tan x}\ , \quad (2.17)$$

and from the triangle NX"X"' we get the following expression:

$$\sin(\lambda - \Omega) = \tan x \cot i\ , \quad (2.18)$$

which is now solved for $\tan x$:

$$\tan x = \sin(\lambda - \Omega)\tan i = \sin\lambda\cos\Omega\tan i - \cos\lambda\sin\Omega\tan i\ . \quad (2.19)$$

Now the angle γ can be eliminated from Eqs. (2.16) and (2.17) to get the following expression:

$$\frac{\sin b}{\cos i}\left(\frac{1}{\sin\beta\cos\beta}\right) + \frac{\tan x}{\sin\beta} = \frac{1}{\cos\beta}\ . \quad (2.20)$$

Finally, by substituting $\tan x$ from Eq. (2.19) we obtain

$$\boxed{X + Y\cos\beta\sin\lambda - Z\cos\beta\cos\lambda = \sin\beta\ ,} \quad (2.21)$$

where the following abbreviations are used: $X = \sin b/\cos i$, $Y = \cos\Omega\tan i$, and $Z = \sin\Omega\tan i$.

Equation (2.21) stems from Delambre (1814) and was extensively discussed by Waldmeier (1955). It has three unknowns and it is necessary to have at least three equations to determine X, Y, and Z. This means that we need at least three observations of a sunspot (or some another object on the Sun) to get one set of values of b, Ω, and i. In this procedure possible small meridional motions of the sunspot are neglected and the heliographic latitude b of a sunspot is assumed to be constant during the disk transit. After X, Y, and Z are observationally determined, the corresponding values of b, Ω, and i can be easily calculated using the following transformation:

$$\boxed{b = \arcsin\left(X\cos i\right)\ ,} \tag{2.22}$$

$$\boxed{\Omega = \arctan\left(Z/Y\right)\ ,} \tag{2.23}$$

$$\boxed{i = \arctan\left(\sqrt{Y^2 + Z^2}\right)\ .} \tag{2.24}$$

Finally, we note that with more than three observations the error of the quantities Ω and i can be calculated, as discussed in Stark and Woehl (1981).

2.2.2 *Coordinate Transformation Based on Linear Algebra and Analytical Geometry*

We now present an alternative derivation of Eq. (2.21) using basic linear algebra and analytical geometry, according to Skokic (2020). Consider an object on the solar surface, for instance a sunspot, having the coordinates λ, β in the celestial heliocentric ecliptic system and coordinates l, b in the heliographic system (the latitude and the longitude, respectively). The vector $\mathbf{r}$ from the center of the Sun to the sunspot on the surface in heliocentric coordinates is defined by

$$\mathbf{r} = \begin{pmatrix} \cos\beta\cos\lambda \\ \cos\beta\sin\lambda \\ \sin\beta \end{pmatrix}. \tag{2.25}$$

Further, the vector $\mathbf{n}$ represents the solar rotation axis. The components of this vector in the heliocentric coordinates is then

$$\mathbf{n} = \begin{pmatrix} \sin i\sin\Omega \\ -\sin i\cos\Omega \\ \cos i \end{pmatrix}. \tag{2.26}$$

While the Sun rotates, an object (e.g., the sunspot) is moving in the plane perpendicular to the solar rotation axis. Here we again neglect meridional motions, as was the case in the previous subsection. This plane is defined with the following equation:

$$\mathbf{n}\cdot\mathbf{r} - d = 0\ , \tag{2.27}$$

where d is the distance of the plane from the center of the Sun:

$$d = \sin b \ . \tag{2.28}$$

We now combine these equations and insert the two vectors to get

$$\sin i \sin \Omega \cos \beta \cos \lambda - \sin i \cos \Omega \cos \beta \sin \lambda + \cos i \sin \beta = \sin b \ . \tag{2.29}$$

As in the previous subsection the same abbreviations are used: $X = \sin b / \cos i$, $Y = \cos \Omega \tan i$, and $Z = \sin \Omega \tan i$, which are substituted into Eq. (2.29), where both sides are divided by $\cos i$. Finally we get

$$\boxed{X + Y \cos \beta \sin \lambda - Z \cos \beta \cos \lambda = \sin \beta \ ,} \tag{2.30}$$

which is identical to Eq. (2.21).

2.2.3 *Measurements of the Solar Rotation Elements*

Solar rotation elements were measured using the method described in this chapter and similar methods, which are more or less approximative. A comprehensive list of results obtained from the seventeenth to twenty-first century is compiled in Table 2.1. The observers and references are included in the table.

Exercises and Questions

1. What is the yearly change of the Ω value?

2. Does the value of Ω increase or decrease with the advancement of time?

3. Calculate the values of Ω for the years 1600, 1700, 1800, 1900, 2000, 2100, and for the year in which you read this book.

4. Draw a sketch of the solar sphere with the vectors and plane mentioned in Sect. 2.2.2.

Table 2.1 Solar rotation elements Ω and i derived from measurements of the rotation of sunspots (s) and using other methods: Doppler (D) and helioseismology (h). All angles are expressed in degrees. The observer Bošković has two entries, since two different methods were applied (for details see the given reference). Note that Ω_{obs} for both entries Balthasar et al. refer to the year 1850

Observer	Year of obs.	Ω_{obs}	Ω_{1900}	Ω_{1976}	i	Method	Source
J. D. Cassini	1678	68	71.1	72.2	7.5	s	Woehl (1978)
J. Cassini	1740	70	72.2	73.3	–	s	Woehl (1978)
La Lande	1775	78.07	79.81	80.87	7.33	s	Woehl (1978)
Delambre	1775	80.12	81.86	82.92	7.32	s	Woehl (1978)
Bošković	1777	70.35	72.07	73.13	7.73	s	Husak et al. (2023)
Bošković	1777	74.05	75.77	76.83	6.82	s	Husak et al. (2023)
Böhm	1833	76.78	77.72	78.78	6.95	s	Woehl (1978)
Laugier	1840	75.13	75.97	77.03	7.15	s	Woehl (1978)
Wiehmann	1846	83.78	84.53	85.59	7.75	s	Woehl (1978)
Carrington	1850	73.67	74.37	75.43	7.25	s	Woehl (1978)
Spoerer	1861/66	74.52	75.07	76.13	6.97	s	Woehl (1978)
Wilsing	1882	75.78	76.03	77.09	7.17	s	Woehl (1978)
Dyson & Maunder	1874–1911	–	74.59	75.65	7.18	s	Dyson and Maunder (1912), Woehl (1978)
Dyson & Maunder	1874–1912	–	74.48	75.54	7.18	s	Dyson and Maunder (1913), Woehl (1978)
Epstein	1905–1910	–	76.30	77.36	7.20	s	Epstein (1916), Woehl (1978)
Epstein	1903–1910	–	73.97	75.03	7.27	s	Epstein (1917), Woehl (1978)
Wöhl	1976	76.31 ± 0.65	–	–	6.77 ± 0.31	D	Woehl (1978)
Stark & Wöhl	1980	75.8 ± 0.27	–	–	7.15 ± 0.034	s	Stark and Woehl (1981)
Balthasar et al.	1874–1984	73.75 ± 0.15	–	–	7.137 ± 0.017	s	Balthasar et al. (1986)
Balthasar et al.	1874–1976	73.86 ± 0.38	–	–	7.12 ± 0.05	s	Balthasar et al. (1987)
Beck & Giles	1996–2001	73.5 ± 0.1	–	–	7.155 ± 0.002	h	Beck and Giles (2005)

References

Balthasar, H., Lustig, G., Woehl, H., Stark, D.: The solar rotation elements i and omega derived from sunspot groups. Astron. Astrophys. **160**(2), 277–279 (1986)

Balthasar, H., Woehl, H., Stark, D.: The solar rotation elements i and Omega derived from recurrent single sunspots. Astron. Astrophys. **174**(1–2), 359 (1987)

Beck, J.G., Giles, P.: Helioseismic determination of the solar rotation axis. Astrophys. J. Lett. **621**(2), L153–L156 (2005)

Carrington, R.C.: Observations of the spots on the sun: From November 9, 1853, to March 24, 1861, made at Redhill (1863)

Cox, A.N. (ed.).: Allen's astrophysical quantities, 4th edn. AIP Press, Springer, New York (2000)

Delambre, J.B.J.: Astronomie theorique et pratique (1814)

Dyson, F.W., Maunder, E.W.: The position of the Sun's axis as determined on photographs of the Sun from 1874–1911. Mon. Not. **72**, 564 (1912)

Dyson, F.W., Maunder, E.W.: Sun, Axis, the position of, from photographs, 1874–1912. Mon. Not. **73**, 673 (1913)

Epstein, T.: Rotationselemente der Sonne. Astronomische Nachrichten **202**(24), 401 (1916)

Epstein, T.: Rotations elemente der Sonne. Astronomische Nachrichten **204**(20), 351 (1917)

Husak, M., Brajša, R., Špoljarić, D., Krajnović, D., Ruždjak, D., Skokić, I., Roša, D., Hržina, D.: Bošković's method for determining the axis and rate of solar rotation by observing sunspots in 1777. Solar Phys. **298**(10), 122 (2023)

Roša, D., Hržina, D., Husak, M., Brajša, R., Špoljarić, D., Skokić, I., Maričić, D., Šterc, F., Romštajn, I.: Determination of the solar rotation elements and period from Ruđer Bošković's sunspot observations in 1777. In: Katya, G., Boian, K., Dimitar, D. (eds). *Proceedings of the Thirteenth Workshop "Solar Influences on the Magnetosphere*, pp. 86–91 (2021)

Scheiner, Christoph: Rosa ursina (1630)

Skokic, I.: Private communication (2020)

Stark, D.: Bestimmung der Rotationselemente Ω und i der Sonne aus Fleckendaten der Jahre 1940 bis 1968. Diplomarbeit, Georg-August-Universität, Göttingen (1979)

Stark, D., Woehl, H.: On the solar rotation elements as determined from sunspot observations. Astron. Astrophys. **93**(1–2), 241–244 (1981)

Waldmeier, M.: Ergebnisse und Probleme der Sonnenforschung, 2nd edn. Akademische Verlagsgesellschaft Geest & Portig K.-G, Leipzig (1955)

Woehl, H.: On the solar rotation elements i and ω as determined by Doppler velocity measurements of the solar plasma. Astron. Astrophys. **62**(1–2), 165–177 (1978)

Chapter 3
Coordinate Systems and Coordinate Transformation

For measurements of the solar rotation a coordinate system on the Sun, which is not a solid body, as for example the Earth, is needed. The Sun consists of plasma. First, the concept of the Carrington rotation is introduced, which represents a frame of reference rotating with the same, constant period. Then the heliographic coordinates, the latitude and the longitude, are defined, in a general analogy with the geographic coordinates. The next step is a procedure for the determination of these heliographic coordinates from observations of solar objects. The main part of this procedure is the transformation of the coordinates measured during observation into the heliographic coordinates. Then the rotation velocity and proper motions of observed structures on the Sun can be determined.

3.1 The Heliographic Coordinate System

Although the topic of this chapter is partially treated in several books, such as Waldmeier (1955), Meeus (1988), Reinsch et al. (1999), Stix (2002), and Hanslmeier (2023), the most comprehensive description with all derivations is given in the book Roša (2020) which is followed here.

3.1.1 Carrington Rotation

For the sunspots at low latitudes R. Carrington introduced in the mid-nineteenth century the mean sidereal rotation period of the Sun amounting to 25.38 days. This rotation period is called the Carrington rotation. The corresponding synodic rotation

R. Brajša and A. Hanslmeier, *Solar Rotation*, UNITEXT for Physics,
https://doi.org/10.1007/978-981-97-6879-0_3

period amounts to 27.2753 days. According to a convention the heliographic longitude increases towards the west on the solar disk. So, the heliographic longitude of the solar disk center decreases with time and during one Carrington synodic rotation period of 27.2753 days it is smaller by 360°. The first Carrington rotation started on November 9, 1853, when the heliographic longitude of the solar disk center was 0°. After one synodic rotation period of 27.2753 days the second Carrington rotation started, and so on. The ordinal number N of any Carrington rotation can be calculated to a very large precision by

$$\boxed{N = 1750 + \text{whole number of} \left[\frac{JD - 2\,445\,871.91}{27.2753} \right] ,} \tag{3.1}$$

where JD denotes the Julian Day. By the whole number of a decimal number it is considered that all digits after the decimal point are neglected.

The heliocentric representation of the celestial sphere is given in Fig. 3.1, where the point K is the ecliptic pole and the point P_0 the north pole of the solar rotation axis. The position of the vernal equinox on the ecliptic is denoted by ♈ and the large circle UNV is the solar equator. The ascending node of the solar equator at the ecliptic is denoted by N. The Sun rotates in the direct sense, which means that viewed from the north pole P_0 the rotation is in the reversed sense from the watch needle, which is indicated by an arrow.

As described in the previous chapter, the position of the solar rotation axis in space is defined by two parameters: the angle between the ecliptic plane and equatorial plane of the Sun (i) and the ecliptic longitude of the ascending node of the solar equator (Ω). The formulae for these parameters are also given in the previous chapter. Note that the angle Ω corresponds to the arc ♈N.

The large circle P_0O in Fig. 3.1 is the prime meridian used for measurement of the heliographic longitude. We assume now that the point O at the solar equator rotates with the Carrington sidereal period of 25.38 days. The reference position of the point O was chosen to coincide with the ascending node (N) of the solar equator on the date January 1, 1854 at 12:00 UT. This corresponds to the Julian Day $JD = 2\,398\,220.0$ and the position of the point O is determined with the angular distance $W =$ NO:

$$\boxed{W = \frac{360^\circ}{25.38\,\text{day}} (JD - 2\,398\,220.0)\ .} \tag{3.2}$$

3.1.2 The Heliographic Latitude and Longitude

We now define the heliographic coordinates of an arbitrary point on the solar disk X as

heliographic latitude B: $B = 90^\circ - P_0X = X'X$, and

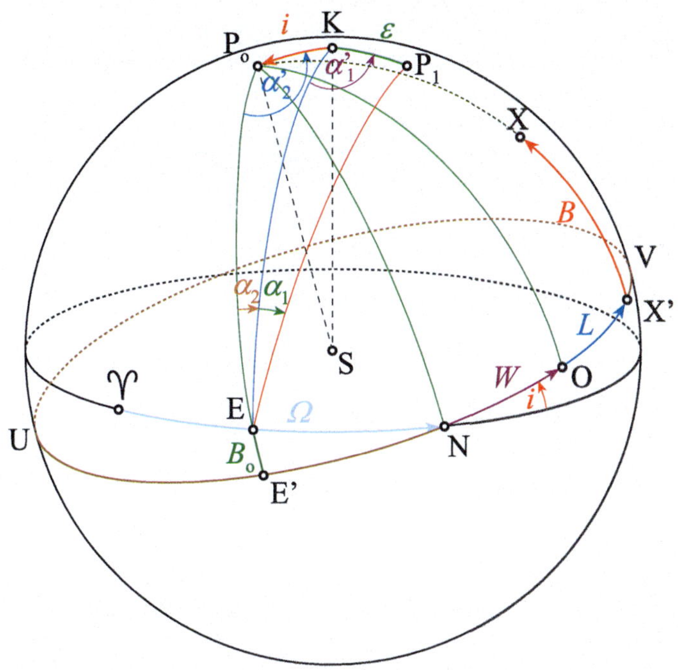

Fig. 3.1 Heliocentric representation of the celestial sphere. Adapted from Roša (2020)

heliographic longitude L: the angle at P_0 of the spherical triangle OP_0X', which is equal to the arc length OX'.

The point E is located at the center of the apparent solar disk projected to the ecliptic and this point lies at the ecliptic. The ecliptic longitude of the point E is equal to $\lambda_0 - 180°$, where λ_0 is the apparent geocentric longitude of the Sun. Further, the following relation is valid:

$$EN = \Omega - (\lambda_0 - 180°) \ . \tag{3.3}$$

We denote the heliographic latitude and longitude of the solar disk center with B_0 and L_0, respectively. Then the following relations hold:

$$P_0E = 90° - B_0 \tag{3.4}$$

$$\sphericalangle EP_0N = 360° - W - L_0 \ , \tag{3.5}$$

where the quantity $360° - W$ denotes the heliographic longitude of the ascending node.

Finally, the rules of spherical trigonometry are applied on the triangles EE'N and P_0EN. We note that in the first case the angle at N is i and in the second case it is $90° - i$. The following relations are valid:

$$\boxed{\sin B_0 = \sin(\lambda_0 - \Omega) \sin i} \tag{3.6}$$

$$\cos B_0 \sin(L_0 + W) = \cos i \sin(\Omega - \lambda_0) \tag{3.7}$$

$$\cos B_0 \cos(L_0 + W) = -\cos(\Omega - \lambda_0) \ . \tag{3.8}$$

By dividing the last two equations we get

$$\boxed{\tan(L_0 + W) = -\frac{\cos i \sin(\Omega - \lambda_0)}{\cos(\Omega - \lambda_0)} \ .} \tag{3.9}$$

For practical calculation, it is important to know to which quadrant belongs the argument $L_0 + W$ of the tangent function at the left-hand side of Eq. (3.9). The heliographic coordinates B_0 and L_0 can be calculated from Eqs. (3.6) and (3.9) knowing the value of the apparent geocentric longitude of the Sun at the time of observation, λ_0.

We now turn back to Fig. 3.1, where we identify the north celestial pole P_1 and the north pole of the solar rotation axis, P_0. The position angle P of the pole P_0 is defined by

$$P = \alpha_1 + \alpha_2 \ . \tag{3.10}$$

The solar position angle P has a positive value when the north pole of the solar rotation axis is turned to the east and a negative value when it is turned towards the west and it changes during the calendar year. Around April 7 the position angle has the lowest value, $P = -26.3°$ and around October 11 the highest value, $P = +26.3°$. Around January 5 and July 7 the position angle has the zero value.

The heliographic latitude of the solar disk center, B_0, tells about the obliquity of the solar rotation axis in respect to the Earth. When the north pole of the solar rotation axis is tilted towards the Earth the value of B_0 is positive and when it is tilted away from the Earth it is negative. Around March 6 the value of this obliquity is $B_0 = -7.25°$ and around September 8 its value is $B_0 = +7.25°$. The value of B_0 is zero around June 6 and December 7.

Finally, we need a formula for the calculation of the position angle P. Referring again to Fig. 3.1, we can see that in the spherical triangle KP_1E the arc KE is equal to 90°, the arc KP_1 is equal to the obliquity of the Earth's equator to the ecliptic, $\varepsilon = 23.44°$, while the spherical triangle α_1' at the point K is equal to the difference of the ecliptic longitude of the point P_1, which amounts to 90°, and the ecliptic longitude of the point E. Expressed mathematically we have

$$\mathrm{KE} = 90^\circ \ , \tag{3.11}$$

$$\mathrm{KP_1} = \varepsilon \ , \tag{3.12}$$

$$\alpha_1' = 90^\circ - (\lambda_0 - 180^\circ) = 270^\circ - \lambda_0 \ . \tag{3.13}$$

Applying rules of spherical trigonometry on the triangle $\mathrm{KP_1E}$ we get

$$\tan\alpha_1 = -\cos\lambda_0 \tan\varepsilon \ . \tag{3.14}$$

In the spherical triangle $\mathrm{KEP_0}$ the arc KE is equal to 90°, while the arc $\mathrm{P_0K}$ is equal to the inclination of the solar equator i. The ecliptic longitude of the point $\mathrm{P_0}$ is equal to $\Omega - 90^\circ$ and the angle α_2' at the point K of the spherical triangle $\mathrm{P_0KE}$ is given by

$$\alpha_2' = \lambda_0 - 180^\circ - (\Omega - 90^\circ) = \lambda_0 - \Omega - 90^\circ \ . \tag{3.15}$$

From the spherical triangle $\mathrm{KEP_0}$ we have the following relation:

$$\tan\alpha_2 = -\tan i \cos(\Omega - \lambda_0) \ . \tag{3.16}$$

We now put Eqs. (3.14) and (3.16) into Eq. (3.10) to finally get the expression for the position angle of the north pole of the solar rotation:

$$\boxed{P = \alpha_1 + \alpha_2 = -\arctan(\cos\lambda_0 \tan\varepsilon) - \arctan[\tan i \cos(\Omega - \lambda_0)] \ .} \tag{3.17}$$

We conclude that the parameters B_0, L_0, and P uniquely define the position of the solar rotation axis in space and the prime meridian, respectively, to the observer at the time of observation using Eqs. (3.6), (3.9), and (3.17).

3.2 Determination of the Heliographic Coordinates From Observations of Objects on the Sun

The correct determination of the heliographic coordinates of the objects observed on the Sun is of a crucial significance for the reliable solar rotation measurements. Even small errors at this stage could lead to false rotation velocity determinations, which might have huge, unwanted implications for the interpretation. So, this topic is rather important and may be very useful for both, the professional and amateur astronomers.

3.2.1 Measured Coordinates on the Observed Solar Disk

The view of the solar disk as seen by the observer is shown in Fig. 3.2a, where the point E is the apparent center of the solar disk. The line EN is perpendicular to the daily path of the Sun and lies along the south–north direction having the position angle equal to zero, $\theta = 0°$. The point P_0 is the north pole of the solar rotation axis and the angle P is the position angle of the north pole of the solar rotation axis. The position of any point X on the solar disk is defined by two quantities which can be determined from observations:

angular distance of the point X from the solar disk center: ρ_1, and

position angle of the point X: $\theta = \angle$NEX.

We now introduce a rectangular coordinate system with the coordinates x and y which can be directly measured from observation of an object at position X. The

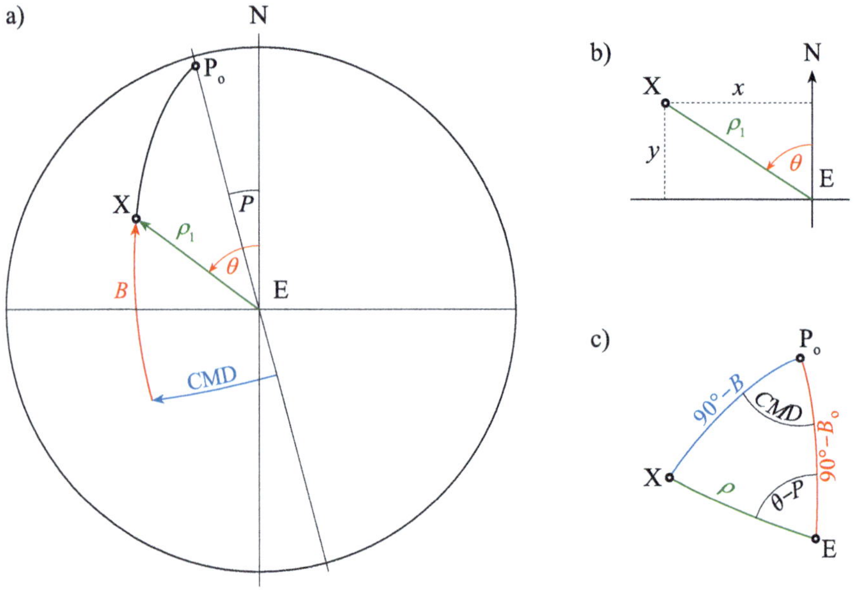

Fig. 3.2 **a** The solar disk as seen by the observer. To determine heliographic coordinates of the observed point X the position angle θ and the angle ρ_1 are determined from observations with the aim to calculate the arc ρ. **b** Determination of the quantities θ and ρ_1 is usually performed by measuring the rectangular coordinates x and y of an identified feature on the solar disk. **c** Relation between the heliographic coordinates L and B and the measured coordinates ρ and θ is established by applying trigonometry on the spherical triangle P_0XE. Adapted from Roša (2020)

origin of this coordinate system is at the point E. The x-axis is in direction of the daily path, while the y-axis is in direction of the zero position angle, which is on the line EN (see Fig. 3.2b). We now express the coordinates x and y in relative units, normalized to the apparent solar radius, and the quantities ρ_1 and θ can be calculated using the following relations:

$$\boxed{\rho_1 = \sqrt{x^2 + y^2}\ ,} \tag{3.18}$$

$$\boxed{x = \rho_1 \sin\theta\ ,} \tag{3.19}$$

$$\boxed{y = \rho_1 \cos\theta\ .} \tag{3.20}$$

3.2.2 *Transformation of the Measured Coordinates into the Heliographic Ones*

By knowing the angular distance ρ_1 and the Earth–Sun distance we can calculate the angular arc length of the large circle EX on the solar surface, ρ. The angle ρ is equal to the angle between the lines from the solar center to the observer and from the solar center to the point X (see Fig. 3.3). For the plane triangle with the vertexes Z (observer on Earth), S (center of the Sun), and X (position of an object on the solar surface) it holds:

$$\boxed{\sin(\rho + \rho_1) = \frac{r \sin\rho_1}{R}\ ,} \tag{3.21}$$

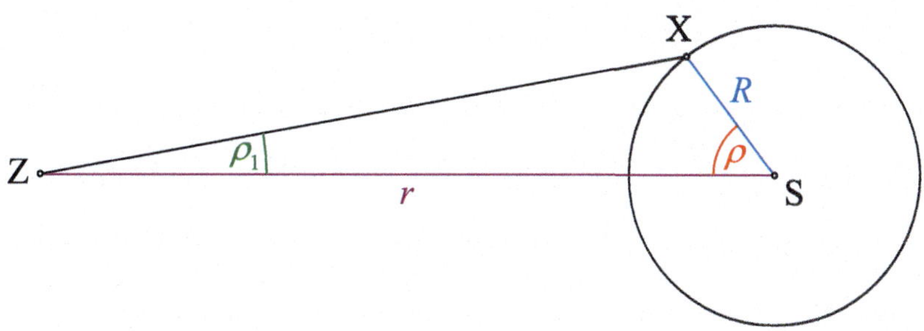

Fig. 3.3 Equation (3.21) can be derived by solving the triangle SZX. The angle at the point X of this triangle is equal to $180° - (\rho + \rho_1)$. Adapted from Roša (2020)

where r represents the Sun–Earth distance and R the solar radius. The quotient of R and r corresponds to the tangent of the angular solar radius. Using Eq. (3.21), the angle ρ can be calculated from the measured angle ρ_1.

The heliographic coordinates of the point on the Sun X which is observed are B and L, while the heliographic coordinates of the solar disk center E are B_0 and L_0. For the spherical triangle P_0XE on Fig. 3.2c the following relations hold:

$$P_0X = 90° - B \ , \tag{3.22}$$

$$P_0E = 90° - B_0 \ , \tag{3.23}$$

$$XE = \rho \ , \tag{3.24}$$

$$P_0EX = \theta - P \ , \tag{3.25}$$

$$XP_0E = L - L_0 \ , \tag{3.26}$$

where the quantity $L - L_0$ denotes the angular distance between the observed point X from the central meridian on the Sun. It is called the central meridian distance and abbreviated by CMD. So, $L - L_0 = CMD$. From the spherical triangle P_0XE it follows:

$$\boxed{\sin B = \sin B_0 \cos\rho + \cos B_0 \sin\rho \cos(P - \theta) \ ,} \tag{3.27}$$

$$\boxed{\cos B \sin(L - L_0) = \sin\rho \sin(P - \theta) \ ,} \tag{3.28}$$

$$\boxed{\cos B \cos(L - L_0) = \cos\rho \cos B_0 - \sin\rho \sin B_0 \cos(P - \theta) \ .} \tag{3.29}$$

With these expressions, the heliographic coordinates B and L of the observed point X can be calculated.

Concluding we can summarize the procedure described in this and the previous chapter. The aim of the procedure is to calculate the heliographic coordinates of an observed object on the solar disk. For the observing time the quantities L_0, B_0, and P should be known as well as the actual Earth–Sun distance r for the same time instant. The angle ρ can be calculated from Eq. (3.21). One way to obtain the correct values of these quantities for any time moment is by using the interpolation of data published in astronomical almanacs. The quantities L_0, B_0, and P are also called the elements for the physical observation of the Sun and they can be directly calculated from Eqs. (3.6), (3.9), and (3.17) if the apparent solar longitude λ_0 and the inclination of the Earth's equator to the ecliptic, ε are known. In this procedure the quantities i, Ω, and W are defined by Eqs. (2.1), (2.2), and (3.2).

Exercises and Questions

1. Describe the concept of Universal Time (UT). In which time zone do you read this book? How can you convert your local time to UT? Take into account possible Daylight Saving Time (DST) in your time zone.

2. Describe the concept of the Julian Day (JD). What JD was at the day when you were born? If you know the precise time of your birth (hour and minute) you can express the JD as a fraction.

3. During which Carrington rotation (the ordinal number) you read this book?

4. Imagine that you observe a sunspot with the projection method (on the screen behind the ocular of the telescope). You measure the position of the sunspot in cm on the paper attached to the screen. Describe the transformation of the coordinates you determined in the local frame of reference into the heliographic coordinates. Identify equations in the text needed for this procedure.

5. Consult the literature about the Porter disk and how it can be used to measure coordinates of solar objects from observations?

References

Hanslmeier, A.: Introduction to Astronomy and Astrophysics. Springer-Varlag, Berlin (2023)
Meeus, J.: Astronomical Formulae for Calculators, 4th edn. Willmann-Bell, Richmond (1988)
Reinsch, K., Beck, R., Hilbrecht, H., Völker, P.: Die Sonne beobachten. Verlag Sterne und Weltraum, Hüthig, Heidelberg (1999)
Roša, D.: Elementarna Astronomija. Croatian Astronomical Society, Zagreb, Zvjezdarnica Zagreb-ZAS (2020)
Stix, M.: The Sun: An Introduction. Springer-Varlag, Berlin (2002)
Waldmeier, M.: Ergebnisse und Probleme der Sonnenforschung, 2nd edn. Akademische Verlagsgesellschaft Geest & Portig K.-G, Leipzig (1955)

Chapter 4
Methods for Solar Rotation Measurements

In this chapter, we describe the methods for measurements of the solar rotation. The method which is most widely used is the tracer method, which is based on following positions of discernible objects in the solar atmosphere during a time span. The apparent positions change due to the solar rotation. Before introducing the tracer method, we give examples of suitable tracers which can be used for position measurements. Then we describe the spectroscopic method (also called the Doppler method), the helioseismic method, and finally the method of correlation of solar images. The advantages and disadvantages of each method will be pointed out and the applicability of each method for specific science cases will be discussed.

4.1 Examples of Suitable Tracers

Almost all discernible structures in the solar atmosphere can be used as tracers for the solar rotation determination. Extensive descriptions of various objects on the Sun can be found in standard textbooks on solar physics, such as Bruzek and Durrant (1977), Zirin (1988), Stix (2002), and Foukal (2013). Here we give three examples of suitable tracers for the solar rotation determination: sunspot and sunspot groups, solar prominences, also called filaments when observed on the solar disk, and coronal bright points.

4.1.1 Sunspots and Sunspot Groups

Sunspots are best visible in the white light images of the solar photosphere (Fig. 4.1). They are darker and cooler structures in comparison to the surrounding photosphere. Sunspots consist of the central part called umbra and the surrounding filamentary part called penumbra. The umbra is darker than the penumbra and their fine structure

R. Brajša and A. Hanslmeier, *Solar Rotation*, UNITEXT for Physics,
https://doi.org/10.1007/978-981-97-6879-0_4

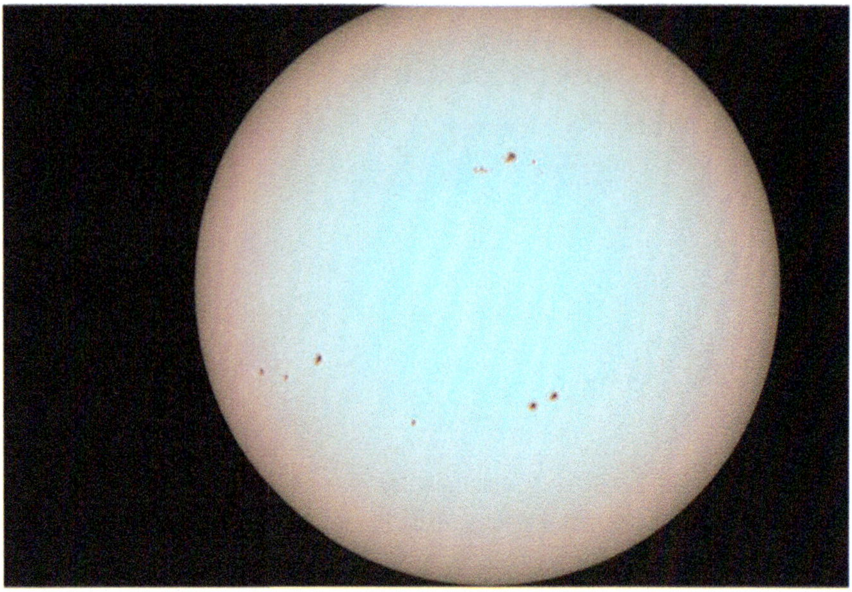

Fig. 4.1 Full-disk solar image taken in white light on August 17, 2023. Several sunspot groups are visible as dark areas in the solar photosphere in the ascending phase of the solar activity. Credit: Arnold Hanslmeier, private observatory

can be clearly seen in, e.g., Fig. 4.2. Sunspots are associated with strong magnetic fields and they often form sunspot groups consisting of two parts having opposite magnetic polarity. Nice examples of large sunspot groups are given in Figs. 4.3, 4.4, and 4.5. Extensive descriptions of the phenomenology and physical properties of sunspots can be found in several textbooks and reviews, e.g., Gurman (1992), Stix (2002), Thomas and Weiss (2008), and Hanslmeier (2023).

Sunspots represent also one of the most evident and important manifestations of the solar activity, which rises and falls in cycles of about 11 years. The relative sunspot number was introduced by R. Wolf in the middle of the nineteenth century, shortly after S. H. Shwabe discovered the 11-year solar activity cycle. The relative sunspot number is defined by a simple formula:

$$\boxed{R = k(10\,g + f)}\ , \tag{4.1}$$

where R is the relative sunspot number, g the number of sunspot groups, and f the total number of all spots. We note that the numbering considers all sunspots and sunspot groups visible on the solar disk and that a single, isolated sunspot is also considered as a sunspot group. The constant k is the calibration factor specific for a particular observatory and telescope, as well as taking into account the local weather and seeing conditions.

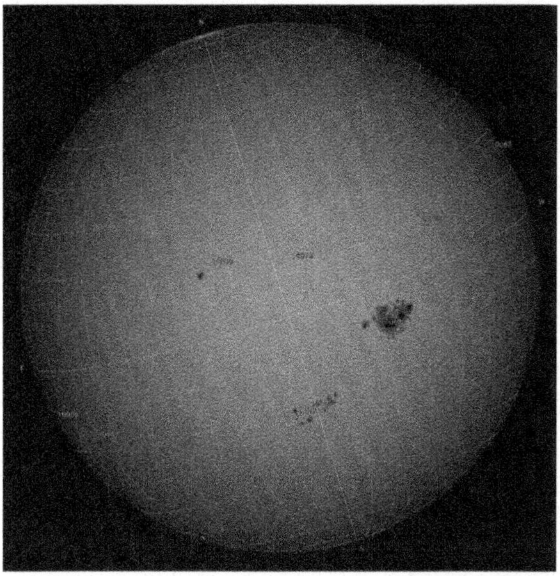

Fig. 4.2 Scanned white light image of the whole solar disk with overlaid grid taken at Kanzelhöhe Observatory for solar and environmental research (Institute of Physics, University of Graz, Austria) on August 19, 2002. Central meridian is represented with the white line connecting the north and the south solar pole. Adapted from Poljančić Beljan et al. (2017)

For the analysis of the solar rotation using sunspots we point out some important properties of the solar activity cycle. A new solar cycle begins with appearance of sunspots at middle to moderately high latitudes on both solar hemispheres, the northern and the southern one. As the cycle progresses new sunspots appear at latitudes closer to the equator on both solar hemispheres. This latitude shift, also known as the butterfly diagram, does not mean that the same sunspots move towards lower latitudes; it just means that the new sunspots emerge and form active regions at lower latitudes, closer to the solar equator. The next solar cycle begins at middle to moderately high latitudes when there are still some of the sunspots belonging to the previous solar cycle present at low latitudes. So, the old and the new solar cycles partly overlap and the term "extended solar cycle" is sometimes used.

So, sunspots can be mostly found at the latitudes from the solar equator up to ± 40 degrees. At the solar equator itself and at the latitudes above ± 40 degrees sunspots appear rarely and they are completely absent from high latitudes and polar regions of the Sun. Moreover, due to the variation of solar activity and the butterfly diagram the number of sunspots is not distributed evenly, neither during the 11-year solar cycle nor over the solar latitudes. There exists immense literature about the solar cycle and here we recommend just a small selection of research papers, reviews, and books: Brajša et al. (2009), Brajša et al. (2022), Clette et al. (2014), Cliver et al. (2013), Cliver et al. (2015), Wilson (1994), Hanslmeier (2020), Harvey (1992), Zirin (1992), Ermolli et al. (2014), and Cliver (2014).

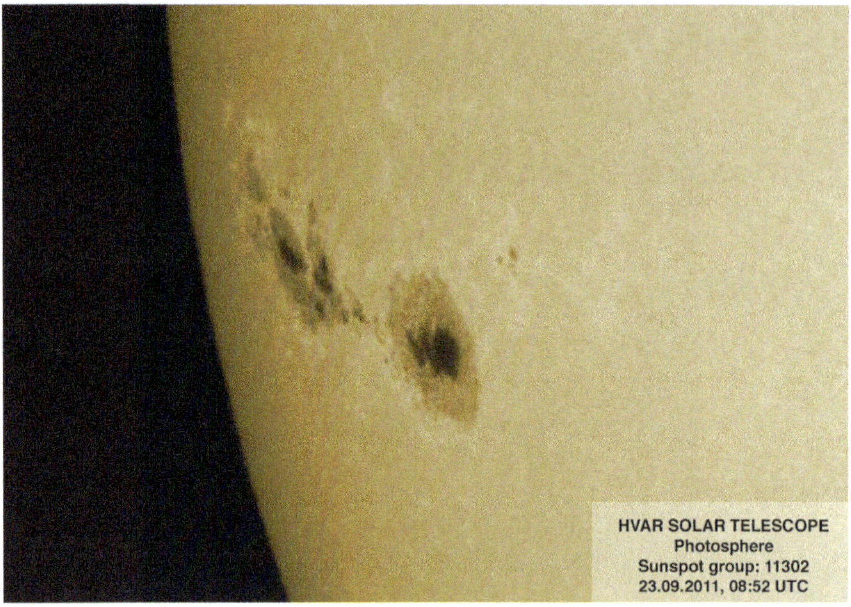

Fig. 4.3 Large sunspot group in the photosphere near the solar limb. The image of the portion of the solar disk was taken with the Hvar Observatory white light solar telescope on September 23, 2011. Credit: Hvar Observatory, Faculty of Geodesy, University of Zagreb, Croatia

Fig. 4.4 Large sunspot group as seen in the image of the portion of the solar disk taken with the Hvar Observatory white light solar telescope on August 24, 2015. Credit: Hvar Observatory, Faculty of Geodesy, University of Zagreb, Croatia

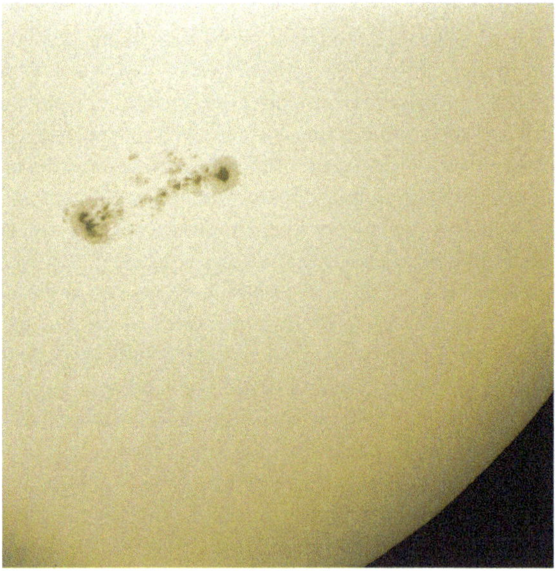

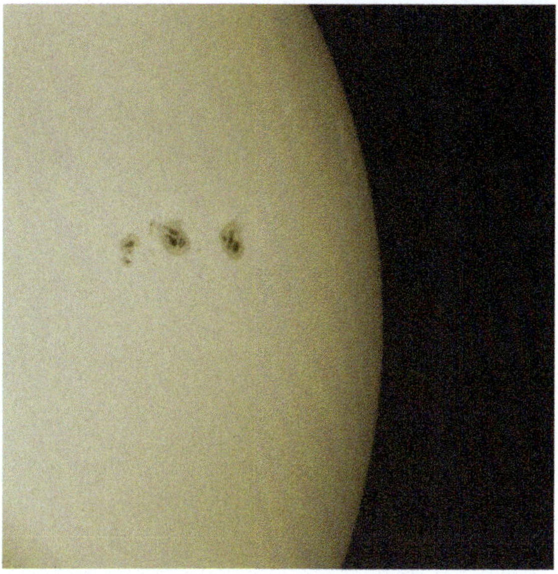

Fig. 4.5 Large sunspot group near the solar limb as seen in the image of the portion of the solar disk taken with the Hvar Observatory white light solar telescope on July 21, 2016. Credit: Hvar Observatory, Faculty of Geodesy, University of Zagreb, Croatia

Sunspots appear in a variety of classes, which are rather different. The classical Zurich classification of sunspots was introduced by M. Waldmeier in the mid-twentieth century and is extensively discussed in his book (Waldmeier (1955)), and later in many standard textbooks, such as Stix (2002) and Hanslmeier (2023). This classification was later modified and there were several attempts in the last few decades to introduce new classifications taking into account improved observing methods and comprehensive physical understanding of the sunspot phenomenon. Here we present two of these alternative sunspot classifications, the McIntosh classification and the Mt Wilson classification, following the paper Dumbović et al. (2024).

Selected examples of the McIntosh sunspot classification are given in Fig. 4.6. *The left panels* show 7 classes of the modified Zurich classification system:

- Class A—unipolar, no penumbrae;
- Class B—bipolar, no penumbrae;
- Class C—bipolar, penumbra on one end of the group;
- Class D—bipolar, penumbra on spots at both ends of the group, length $< 10°$;
- Class E—bipolar, penumbra on spots at both ends of the group, $10° <$ length $< 15°$;
- Class F—bipolar, penumbra on spots at both ends of the group, length $> 15°$;
- Class H—unipolar, penumbra.

The middle panels show 6 classes of the penumbra of the largest sunspot:

- Class x—no penumbra;
- Class r—rudimentary penumbra;

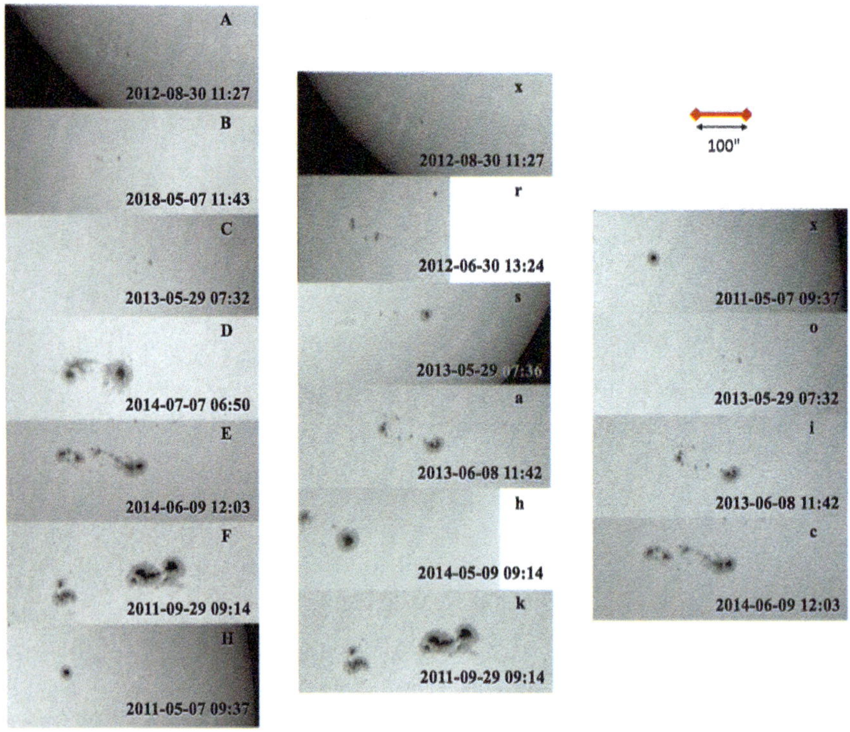

Fig. 4.6 Examples of sunspot groups taken using white light observations at Hvar Observatory to illustrate the McIntosh sunspot classification. The times of observations are given at each panel and for the details see text. Adapted from Dumbović et al. (2024). Credit: Hvar Observatory, Faculty of Geodesy, University of Zagreb, Croatia

- Class s—small, symmetric penumbra;
- Class a—small, asymmetric penumbra;
- Class h—large, symmetric penumbra;
- Class k—large, asymmetric penumbra.

The right panels show 4 classes of the sunspot distribution:

- Class x—undefined relative spottedness (for unipolar groups);
- Class o—open spottedness (few and small interior spots);
- Class i—intermediate spottedness (numerous interior spots, but without penumbra);
- Class c—compact spottedness (numerous interior spots, some with penumbra).

Selected examples of the Mt Wilson sunspot classification are given in Fig. 4.7. *The left panels* show examples from white light observations, while *the right panels* present corresponding SDO/HMI images showing the magnetic field of opposite polarity denoted by black and white colors. Top-to-bottom panels show the following:

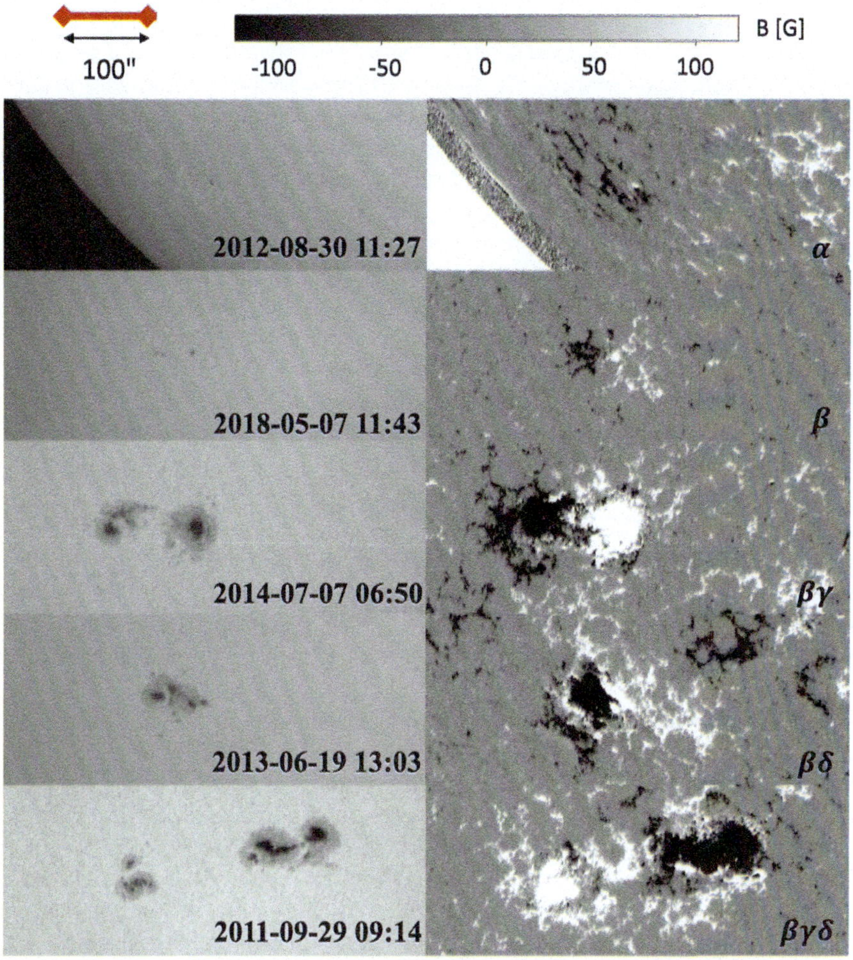

Fig. 4.7 Examples of sunspot groups obtained using white light observations at Hvar Observatory (left panels) and corresponding solar SDO/HMI magnetograms (right panels) to describe the Mt Wilson sunspot classification. The times of observations are given at each panel and for the details see text. Adapted from Dumbović et al. (2024). Credit: Hvar Observatory, Faculty of Geodesy, University of Zagreb, Croatia and NASA SDO/HMI

- Class α—unipolar;
- Class β—simple bipolar;
- Class $\beta\gamma$—complex bipolar;
- Class $\beta\delta$ —Class β with a δ spot;
- Class $\beta\gamma\delta$ —Class $\beta\gamma$ with a δ spot.

We note that a δ spot (or δ region) marks a region of closely placed opposite polarity umbrae within a common penumbra.

The main advantage of using sunspots as tracers for the solar rotation determination is their visibility and detectability on the solar disk. Moreover, a very large number of professional and amateur astronomers have been observing sunspots since their telescopic discovery in the early seventeenth century. A very wide geographical distribution of solar patrol stations assures continuous solar observations, so that any sunspot present on the solar disk could barely remain undetected. In total, we have now more than 400 years of continuous sunspot observations, starting with visual observations, followed by the photographic detection, and finally using the CCD devices mounted either on terrestrial telescopes or the ones placed on the satellites. There are many sunspot data sets spanning from several decades to several centuries in which the heliographic positions of sunspots and sunspot groups are listed in a standardized way. These homogeneous data sets enable wide usage of sunspots as favorable tracers for the solar rotation determination, including investigation of possible long-term variations. Also, the recent research on historic sunspot observations fills the gaps in the existing data sets, minimizing the time intervals without any sunspot observations. Finally, an important advantage of using sunspots as tracers is the fact that many sunspot groups live over days, weeks, and even months. A transition of a large sunspot group over the solar disk can be seen in Fig. 4.8. Some sunspot groups can be identified when they reappear after passing over the invisible (back) part of the Sun. The long-living sunspot groups can be traced during several solar rotations, enabling reduction procedures not possible for most other available tracers.

On the other hand, there are unfortunately quite a few disadvantages of using sunspots and sunspot groups for the solar rotation determination. We mention here the three most important disadvantages of sunspots. First, sunspots occupy only low to middle latitudes. This means that sunspots carry a little information about the equatorial solar rotation and absolutely no information about the rotation at high latitudes and in polar regions. Moreover, due to the prominent butterfly diagram mentioned earlier, in a specific phase of the solar cycle, we can observe sunspots only in a relative narrow latitude belt. This means that in different phases of the solar cycle we actually observe sunspots at different latitudes. The second problem is that sunspot groups appear in a variety of sizes and shapes, from a single, round-shaped sunspot to the large, complex groups. In the former case, the coordinate determination is relatively easy, but in the latter case, the position determination of a complex sunspot group is not an easy task. We will later see how this problem can be overcome. The third important disadvantage of sunspots is their proper motions and evolution of sunspot groups. This means that the measured change of the position of sunspots and sunspot groups is not only a consequence of the solar rotation, but also of their displacements within the rotational frame of reference.

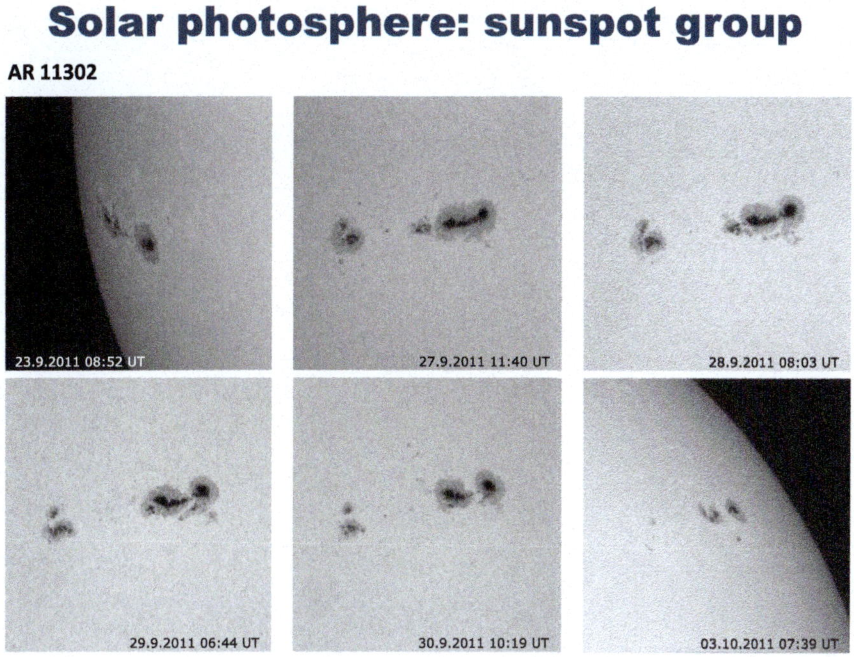

Fig. 4.8 Passage of a large sunspot group over the solar disk during solar rotation. Images were taken with the white light solar telescope at Hvar Observatory on the dates/times given on each panel. Credit: Hvar Observatory, Faculty of Geodesy, University of Zagreb, Croatia

4.1.2 Hα Filaments

Hα filaments and prominences are the two names of the same physical object in the solar atmosphere, namely the solar prominences. They can be observed in the light of many spectral lines, most notably in the hydrogen Hα line. When seen at solar limb, just outside the solar disk, they are seen in emission (and are called simply prominences), while on the solar disk they are seen in absorption (and are then called Hα filaments).

Several illustrative examples of Hα filaments are presented in Figs. 4.9, 4.10, 4.11, and 4.12. In Fig. 4.10, we can see a prominence at the solar limb, having one of its part at the solar disk and the other just outside the solar limb. The part of the prominence outside the limb is bright and the part on the disk is dark against the background in the light of Hα line. As already mentioned, the first case can be interpreted as emission and the second case as absorption. The appearance of dark Hα filaments on the solar disk can be understood using the following simple concept. The prominence which is stretched from the chromosphere into the corona absorbs the light from the photosphere. The transferred energy is then reemitted in all directions, which explains the radiation deficit in the direction of the observer.

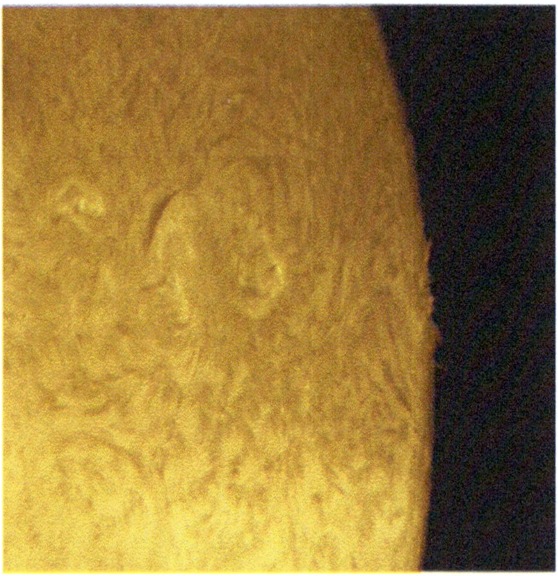

Fig. 4.9 A small filament near the solar limb recorded on November 1, 2011 with the Hvar Observatory Hα solar telescope. Credit: Hvar Observatory, Faculty of Geodesy, University of Zagreb, Croatia

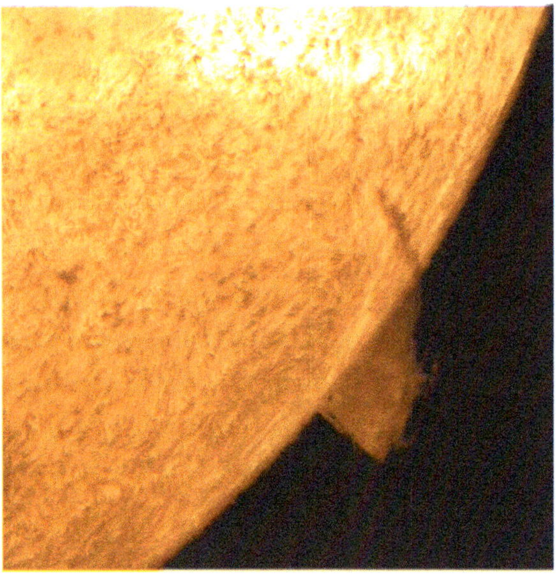

Fig. 4.10 A large filament at the solar limb recorded on August 28, 2012 with the Hvar Observatory Hα solar telescope. It can be clearly seen that the same structure is observed in emission outside the solar limb (as a bright prominence) and in absorption on the solar disk (as a dark filament). Credit: Hvar Observatory, Faculty of Geodesy, University of Zagreb, Croatia

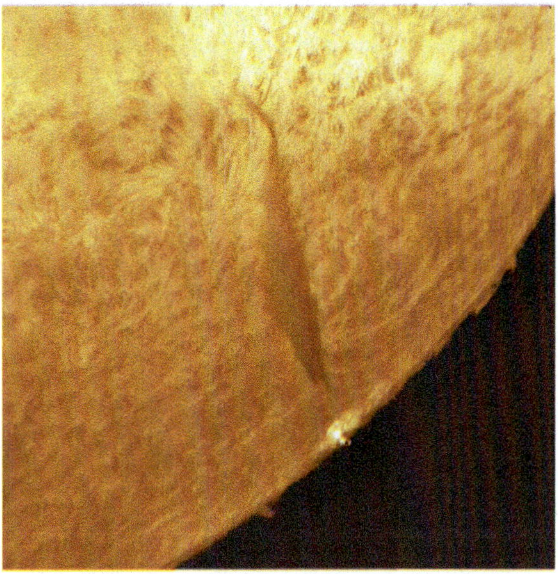

Fig. 4.11 A large filament near the solar limb recorded on August 30, 2012 with the Hvar Observatory Hα solar telescope. Note the footpoints of the filament anchored in the chromosphere. Credit: Hvar Observatory, Faculty of Geodesy, University of Zagreb, Croatia

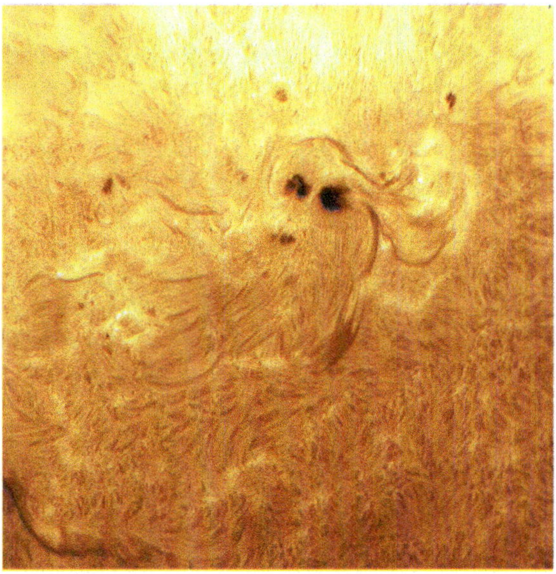

Fig. 4.12 A group of several active region filaments recorded on July 11, 2012 with the Hvar Observatory Hα solar telescope. For details see text. Credit: Hvar Observatory, Faculty of Geodesy, University of Zagreb, Croatia

Prominences can be generally divided into two classes: quiescent prominences and active region prominences (see, e.g., Zirin (1988)). Quiescent prominences appear outside of active regions and they are long-lived. Active region prominences occur in active regions (a nice example can be seen in Fig. 4.12) and they are often related to solar flares and are in fact transient phenomena. Typical dimensions of quiescent prominences are the length: 60 000–600 000 km, the height: 15 000–100 000 km, and the thickness: 5000–15 000 km (e.g., Tandberg-Hanssen (1995)). Almost all prominences lie along the so-called neutral magnetic lines, which separate the adjacent, unipolar magnetic fields of opposite polarity. They are anchored in the photosphere and are extended in the chromosphere and corona.

Observations and models of prominences are extensively reviewed in several papers and books: Tandberg-Hanssen (1995), Ruzdjak and Tandberg-Hanssen (1990), Parenti (2014), and Gibson (2018). The plasma density of prominences is about 100 times higher and the temperature is about 100 lower than the surrounding corona. These factors are rather approximate and can be significantly different from the mentioned values. If they are equal then there is a hydrostatic equilibrium between the prominence and the ambient plasma. If not, then the pressure is balanced through the magnetic pressure. Finally, we come to the question of prominence support against gravity, which is in all cases due to the magnetic field. In a very simplified picture, we can visualize a prominence as a filament of colder and denser plasma suspended in the magnetic arcade connecting photospheric regions of opposite polarity.

As in the case of sunspots, we finally briefly discuss the advantages and disadvantages of using prominences, i.e., Hα filaments as tracers for the solar rotation determination. The main advantage of Hα filaments is that they are relatively stable, long-living structures with good defined edges. These properties enable reliable position measurements of some filament parts. Moreover, prominences can be in principle found at all solar latitudes, which is an important advantage in comparison to sunspots. There is a special subtype of solar prominences, called polar crown filaments, which are formed at high solar latitudes encircling the solar poles. This enables solar rotation measurements in the polar regions of the Sun.

On the other hand, the observers face also some disadvantages of using Hα filaments as tracers for the solar rotation determination. The main challenge is a large and extended structure of solar prominences, especially the quiescent ones. Moreover, because of their pronounced 3D structure, observers measure in fact the positions of filaments parts projected on the solar disk. This is the cause of projection effects which not only result in displaced apparent positions of the measured positions, but this displacement changes during the passage of a filament across the solar disk. Of course, the wrong position measurements can yield false rotation velocities and to a lesser extent false position determination (latitude and longitude) of the tracers. These unwanted effects can be overcome in two ways. First, we can select the footpoints of the Hα filaments as tracers. In this way, the position measurements are confined to well-defined objects, which are also lower in the solar atmosphere which surely minimizes the projection effects. The second possibility, which does not exclude the first one, is an application of the height correction during data reduction. In the next chapter, we will describe the height correction procedure in detail, which leads to the corrected rotation velocity and position of the measured tracers. We finally mention the latitudinal distribution of Hα filaments. The coverage of the solar disk is surely better than for the sunspots, but there are still preferred latitude belts for prominences during the solar cycle. So, the latitudinal coverage is not complete for a chosen time interval.

4.1.3 Coronal Bright Points

Coronal bright points are omnipresent activity phenomena in the solar atmosphere. They consist of low-corona small-scale plasma loops that connect two magnetic flux concentrations of opposite polarities in the photosphere (Madjarska (2019) and references therein). They were first observed with the series of rocket missions in the late 1960s and early 1970s, as well as with the solar instruments onboard the Skylab mission, and were detected as small point-like features of bright soft X-ray emission. So, they were originally called X-ray bright points. For the details see Golub and Pasachoff (2009) and Golub et al. (1974). The satellite Yohkoh provided later improved soft X-ray images of the Sun in which the X-ray bright points could be studied in more detail. Later with the launch of several space missions, such as the Solar and Heliospheric Observatory (SOHO) and the Solar Dynamics Observatory (SDO), it became possible to observe the Sun in UV and EUV light, and so the small bright points, visible in those images, were called EUV bright points instead of X-ray bright points. Nowadays, the term coronal bright points has been adopted for this phenomenon.

The formation and evolution of coronal bright points are associated with newly emerging magnetic flux, called ephemeral regions, or with the random encounters and canceling of the converging magnetic flux. Here we are speaking about the proper motions of magnetic elements on the solar surface. Dimensions of coronal bright points are in the range from 5 arc sec to about 50 arc sec and their lifetimes are on the scale of hours and days. Coronal bright points can be found in the quiet Sun regions, within and outside coronal holes, and in the vicinity of active regions. More details about their morphology and statistics can be found in a recent study (Matković et al. (2023) and references therein).

In Fig. 4.13 we can see a nice example of identified coronal bright points in a EUV image of the full solar disk Sudar et al. (2015). It is important to note a very good spatial distribution of the tracers. More details about the spatial distribution of coronal bright points can be found in, e.g., Brajša et al. (2005). Although all coronal bright points consist of loop systems of hot and dense confined plasma, some morphological differences could be observed, depending on the spatial resolution of the images. So, a simple phenomenological classification was introduced, dividing coronal bright points into three classes according to their observational appearance Brajša et al. (2002): the point-like structures, the small loops, and the small active regions. We

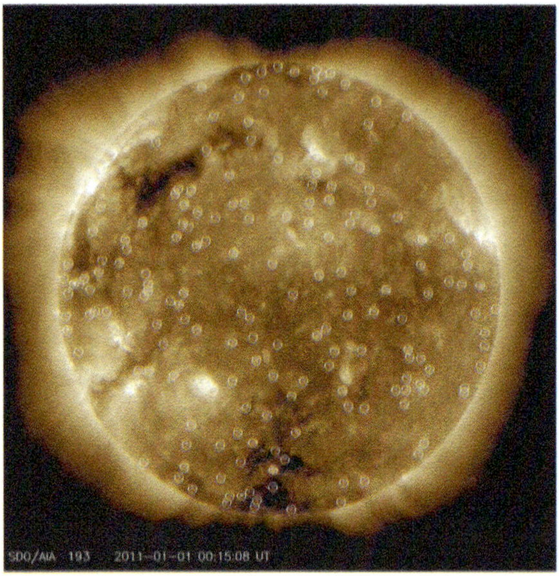

Fig. 4.13 Full-disk EUV image of the Sun taken in the 19.3 nm channel with the SDO/AIA instrument on January 1, 2011. White circles show detected coronal bright points in this image. Adapted from Sudar et al. (2015)

can conclude that coronal bright points are in fact downscaled active regions, but spread over the whole solar atmosphere.

As in the case of sunspots and prominences, we now briefly discuss some advantages and disadvantages of coronal bright points when they are used as tracers for the solar rotation determination. Based on the fundamental characteristics of coronal bright points described above, it is easy to list their main advantages. Coronal bright points can be observed at practically all positions on the solar disk and they are present in all phases of the solar cycle. They are localized and relatively stable objects which make them very convenient tracers which are always related to the underlying small-scale magnetic structures.

On the other hand, the observers are confronted also with some disadvantages of coronal bright points when they are used as tracers. The first issue is the relative short lifetime of these tracers, which is, as we have seen, on the scale of hours and days. This difficulty is being overcome with the significantly improved cadence of the solar images obtained with the EUV instruments mounted on satellites recently used. As an example we note that the EIT instrument onboard the SOHO mission in a regular mode produced four full-disk solar images per day, i.e., one image every 6 hours Brajša et al. (2001). The AIA instrument mounted on the SDO satellite, the successor of SOHO, has the cadence of full-disk solar images of 10 minutes Sudar et al. (2015), which is a rather huge leap forward regarding the temporal resolution. Moreover, the spatial resolution was also significantly improved with SDO/AIA in comparison to SOHO/EIT. The second disadvantage is projection effects. Similarly to prominences, coronal bright points also extend to the solar corona, although to much lower heights, but still not negligible. In the next chapter, we will see how this influence, which could lead to false rotation rate determinations, can be overcome. Finally, the third issue is the fact that coronal bright points can be hardly seen in active regions, because of the strong EUV emission of plasma above active regions, which is also from the regions higher than the tops of coronal bright points Brajša et al. (2021).

4.2 The Tracer Method

The tracer method for the solar rotation determination is based on measuring the apparent position of objects on the solar disk, which is changing due to the solar rotation. By following these objects, the so-called tracers, their positions can be measured, typically on a daily basis. By dividing the change in position with the elapsed time, the synodic angular rotation velocity can be calculated.

One of the first astronomers who systematically observed the Sun and applied the tracer method in the early seventeenth century was Christoph Scheiner. An example of his observations is presented in Fig. 4.14 taken from his famous book Rosa Ursina Scheiner (1630). Scheiner used the projection method for position measurements of sunspots, as can be seen in Fig. 1.6. Scheiner performed his solar observations with remarkable accuracy and his extensive observational data set was recently reduced again and prepared for modern analysis Arlt et al. (2016). The high precision of Scheiner's measurements can be illustrated by the fact that they are used in many contemporary investigations of the Sun, including the determination of the solar rotation in the early seventeenth century Sudar and Brajša (2022).

As we have seen the tracer method relies on following solar structures as they are moving across the solar disk due to the solar rotation. A nice example of a large sunspot group passing across the solar disk can be seen in Fig. 4.8.

All methods for the solar rotation determination have their advantages and disadvantages and this is also the case for the tracer method. The most important advantage of the tracer method is the fact that the observer mostly can correctly identify (often visually) the one and the same object in the sequence from one image to the next one. This is the critical point, since misidentifications of tracers represent one of the most serious sources of errors in rotation measurements. Another important advantage is a broad selection of available tracers, covering various spectral ranges, latitudinal coverage, and layers in the solar atmosphere. Finally, for some widely used tracers, notably sunspots, there are plenty of measurements, from many observatories, and available data are extending to the last four centuries.

Unfortunately, there are also quite a few disadvantages of the tracer method for the solar rotation determination. We list the most important ones, roughly based on the review Schroeter (1985), with some additions and updates:

- Intrinsic proper motions caused by the changes within the magnetic field pattern: All tracers are magnetic elements and the source region of plasma can change the

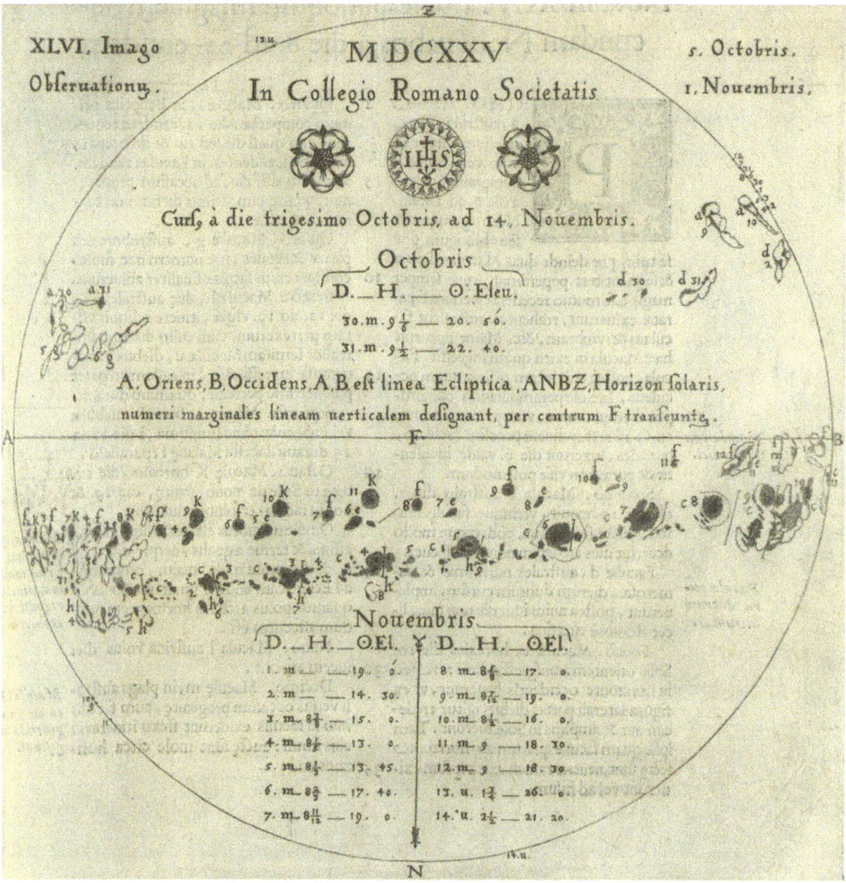

Fig. 4.14 Sunspot drawing made by Christoph Scheiner in his book Rosa Ursina Scheiner (1630). Credit: Wikipedia

position within the flux tube in the time between successive measurements leading to a shift in the position measurement not related to the solar rotation.

- Evolution of a structure which is used as the tracer: This is basically an effect of the proper motion of the object which again is not directly related to the solar rotation. Like in the previous case local magnetic activity is the main cause leading again to the unwanted change in the position measurement.
- Projection effects: For tracers extended to the chromosphere and corona their projection to the photospheric level is in fact measured, which can lead to the false rotation results. An opposite effect is the Wilson depression for the sunspots and faculae, which similarly can result in wrong position measurements close to the solar limb. This unwanted effect could be overcome with various procedures, some of which will be described in detail in the next chapter of this book.

- Differential refraction of the Earth's atmosphere: This effect can lead to distortion of the apparent solar disk and to the false coordinate determination.
- Locally variable distortion of the image scale due to the atmospheric seeing and weather effects.
- Instrumental effects such as telescopic aberrations.
- Inaccuracy of positioning of the solar axis in space (the Carrington elements P and B_0): As we have seen in previous chapters these values are used in the coordinate transformation and even small errors of these quantities can lead to false coordinate of the tracers and consequently to wrong rotation velocities.
- the Butterfly diagram: Almost all tracers used for the solar rotation determination show some kind of preferred latitudes depending on the phase of the solar cycle. This means that in a specific observing time interval, say one year, we do not have a complete latitudinal coverage of the Sun for a used tracer. However, there are some tracers for which this effect is rather small, such as coronal bright points discussed earlier in this chapter.
- Personal errors, especially when solar observers change.

The tracer method can be applied in three ways, as the daily shift method, as the passage method, and as the period method. We describe these three cases in the next three subsections.

4.2.1 The Daily Shift Method

The daily shift method is the simplest and most straightforward application of the tracer method. The longitudinal position of a tracer is expressed as the Central Meridian Distance (CMD), which is measured in the east and west direction from the central meridian; see Fig. 4.2. Synodic rotation velocity ω_{syn} is determined by dividing the change in the CMD by the elapsed time t from two consecutive solar images:

$$\boxed{\omega_{\text{syn}} = \frac{\Delta\, CMD}{\Delta\, t}\,.} \tag{4.2}$$

The positions and times are measured in successive images and the increments are simply calculated by their differences: $\Delta\, CMD = CMD_2 - CMD_1$ and $\Delta\, t = t_2 - t_1$, where the indices 1 and 2 refer to the first and the second measurement, respectively. The measured rotation velocity is usually expressed in degrees per day, although other units are also in use, as will be described later.

The daily shift method is one of the earliest procedures for the solar rotation measurement and it was typically applied to the daily drawings (and later photographs) of the whole solar disk. This explains also the name of the method, although later it was used for any pairs of tracer position measurements. With the development of observing techniques the cadence and precision of solar observations significantly increased. For instance, a typical cadence of solar EUV images with the SOHO satellite is 4 images per 24 hours and with the SDO satellite one image every 10 minutes. We see how dramatic is this improvement compared to the cadence of one solar image per day, but the same method can be used for the pairs of consecutive images of any cadence.

As with any of the methods for the solar rotation determination, the daily shift method has its advantages and disadvantages. The main advantage of this method is the reliable and unambiguous identification of the chosen tracer. Every experienced observer can identify the tracer from one image to the other simply by visual inspection. In this way, the errors caused by the misidentification of tracers are minimized. Further, this method provides the largest possible number of individual rotation velocities measured in a series of images. So, a comprehensive statistical analysis is possible. Finally, the objects which are present in only two consecutive images are included in the analysis, which is not the case in the other subtypes of the tracer method, as will be described later.

There are, however, two disadvantages of the method which cannot be neglected. The first one is the large influence of the proper motion of the tracers, not related directly to the solar rotation. If such proper motions are large, position measurements are significantly influenced and an extremely large or small rotation velocity could be "measured", which is an obvious error. The second disadvantage is that there is no possibility to determine/estimate an error in individual rotation velocity measurement.

The daily shift method is widely used in the solar rotation measurements. Examples of successful application of this method can be found in the following papers (the list is of course not exhausting): Dupree and Henze (1972), Balthasar and Woehl (1980), Howard et al. (1984), Balthasar et al. (1986), Kambry and Nishikawa (1990), Howard (1990), Howard et al. (1990), Zappala and Zuccarello (1991), Sivaraman et al. (1993), Godoli et al. (1993), Javaraiah and Gokhale (1997), Meunier et al. (1997), Suzuki (1998), Ruždjak et al. (2004), Brajša et al. (2006), Sudar et al. (2014), Poljančić Beljan et al. (2017), Sudar et al. (2017), Ruzdjak et al. (2017), Ruzdjak et al. (2018), Poljančić et al. (2022), Ruždjak et al. (2023), and Husak et al. (2023).

4.2.2 The Simple Passage Method

The passage method as a subtype of the tracer method consists of simply following an identified tracer on the Sun in at least three consecutive solar images. In earlier (up to the second half of the twentieth century) applications of this method typically daily images of the Sun were used. Often one or two days were omitted due to the bad weather or other reasons when the observations could not have been performed. However, this is not a serious issue if the same tracer was correctly identified in the next usable image. In this way, a series of central meridian distance (CMD) values with corresponding observation times (t) is obtained: $CMD_1(t_1), CMD_2(t_2), CMD_3(t_3), \ldots CMD_n(t_n)$. Then the linear least-square fit is calculated from the observed data and the synodic rotation velocity is simply the slope of the best-fit line $CMD(t)$.

An example of the function $CMD(t)$ is presented in Fig. 4.15. Three objects in the radio maps of the Sun measured at the frequency of 37 GHz (corresponding approximately to the wavelength of 8 mm) were followed during several days as they passed across the solar disk Pohjolainen et al. (1991).

Modern instruments for solar observations, either terrestrial or mounted on the satellites, enable a much higher cadence of measurements than one image per day. Dozens and even hundreds of solar images per day become available and the passage method can be successfully applied. We note that the names of the methods, the daily shift and the passage method, have historic origins from the times when one solar image per day was usual. As we have seen both methods can now be applied to high cadence data sets obtained within a day or even a fraction of the day.

In addition to the usual least-square method for the fitting of the function $CMD(t)$ newer adaptations of the method are used, such as the robust linear least-squares fit. For details, see for example Poljančić et al. (2022) and references therein.

The main advantage of the simple passage method is that more than two measurements should yield a more reliable final result, i.e., the rotation velocity. Moreover, an application of the linear least-square method enables obtaining the error of the determined rotation velocity. This can be used as a quality test for the measurements. In this way, rotation velocity results with large errors could be eliminated and a systematic criterion could be established.

On the other hand, some disadvantages are also associated with this method. The most important issue is the large influence of the erroneous single position measurements in the $CMD(t)$ series on the resulting least-square fit. The second issue of the simple passage method is that all calculated rotation velocity values are counted with the same statistical weights, regardless of the number of position measurements.

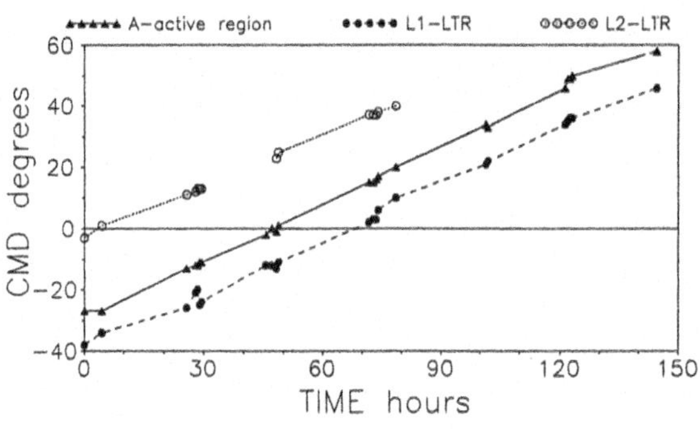

Fig. 4.15 Central meridian distance (CMD) as a function of time. Positions of active regions (AR) and low temperature regions (LTR) were performed on full-disk solar images taken with the 14 meter Metsähovi radio telescope, Helsinki University of Technology, Finland. Adapted from Pohjolainen et al. (1991)

The obstacle can be overcome using the passage method with statistical weights, as we will see in the next subsection.

There are several examples of successful application of the simple passage method. As in the previous subsection, the list is in no way exhausting: Liu and Kundu (1976), Balthasar et al. (1982), Woehl (1983), Dezso and Kovacs (1994), Brajša et al. (2001), Brajša et al. (2002), Brajša et al. (2009), Wöhl et al. (2010), Sudar et al. (2015), Poljančić Beljan et al. (2017), and Poljančić et al. (2022).

4.2.3 The Passage Method with Statistical Weights

The passage method for the determination of the solar rotation can be extended and improved by introducing the statistical weights procedure. Statistical weights (SW) are assigned to the measured rotation velocities according to the duration of the tracing, e.g., two consecutive images: $SW = 1$, three consecutive images: $SW = 2$, four consecutive images: $SW = 3$, etc. So, $SW = d - 1$, where d stands for the number of images. We note that originally one image per day was used, so that d was in fact the number of tracing days. In application of the method the measured rotation velocities are counted SW times. So, the effective number of velocity values in the analysis is given by

$$n_{SW} = \sum_{j=1}^{n} (SW)_j \ , \tag{4.3}$$

where $(SW)_j$ represents the statistical weight of the rotation velocity value for the jth identified tracer and n is the number of identified tracers in a group of measurements. The assumption $SW = d - 1$ is used to have the effective number of resulting velocity values equal to the number of the velocity values, which could be obtained from pairs of two consecutive images (i.e., using the daily shift method).

The passage method with statistical weights was earlier used in the papers Bruzek (1961) and Brajša et al. (1991), while the method was later developed in detail in Brajša et al. (2000), where also several special cases were discussed. We can conclude that the main advantage of the passage method with statistical weights is in the fact that the velocity values, which are obtained with more individual position measurements, are also given more statistical weights. This is justified by the reasonable assumption that rotation velocities calculated from more input data are also more precise.

4.2.4 The Period Method

The last subclass of the tracer method, which we discuss here, is the period method. The main criterion for the applicability of this method is that the selected tracer lives long enough to be correctly identified in two subsequent solar rotations and that in each passage the central meridian was crossed. The central meridian distances of such tracers are fitted by the linear least-square method as a function of time for each passage. In this way, the times of the consecutive passages of the tracer through the central meridian, t_{01} and t_{02}, are calculated. The synodic rotation period of the Sun is then determined as the difference of these two times:

$$\boxed{P_{\text{syn}} = t_{02} - t_{01}\ .} \tag{4.4}$$

From the synodic period the synodic rotation velocity can be easily calculated.

A good example of the adequate tracer for the period method is the long-living recurrent sunspots, in particular belonging to the modified Zurich-type H (more details about the sunspot classification can be found earlier in this chapter). Successful application of this method can be found in Brajša et al. (2002), where solar differential rotation was determined using stable recurrent sunspot groups from the Greenwich data set. We note finally that the main advantage of the period method is a very high precision of the measured rotation rate. On the other hand, an obvious disadvantage of the method is the lack of suitable long-living recurrent tracers.

4.3 The Doppler Method

The Doppler or spectroscopic method is based on the Doppler effect, which will be described in the next subsection. The solar rotation velocity is determined by measuring Doppler shifts of photospheric and more recently chromospheric spectral lines. It is important to know the optical depth of the used spectral line and the height range of the line formation. The main assumption of the method is that the part of the Sun moving towards the observer is blue shifted and the part of the Sun moving away is red shifted; see Fig. 4.16.

The Doppler method is described in detail in several reviews on solar rotation, e.g., in Stix (1989), Howard (1984), Beck (2000), Schroeter (1985), Snodgrass (1992), and Wöhl (1990). As with any method for the solar rotation determination, the Doppler method also has its advantages and disadvantages. The principal advantage of the Doppler method is that it can measure rotation rates at almost all solar latitudes and in all solar cycle phases. The method, however, becomes more and more unreliable as we approach the solar poles.

Similarly to all methods for the solar rotation measurement, the Doppler method has also several important disadvantages. Here we discuss the issues listed in Schroeter (1985) with some additional comments from Howard (1984) and Beck (2000):

- influence of granular motions;
- influence of supergranular motions;
- influence of solar oscillations;
- influence of torsional oscillations;

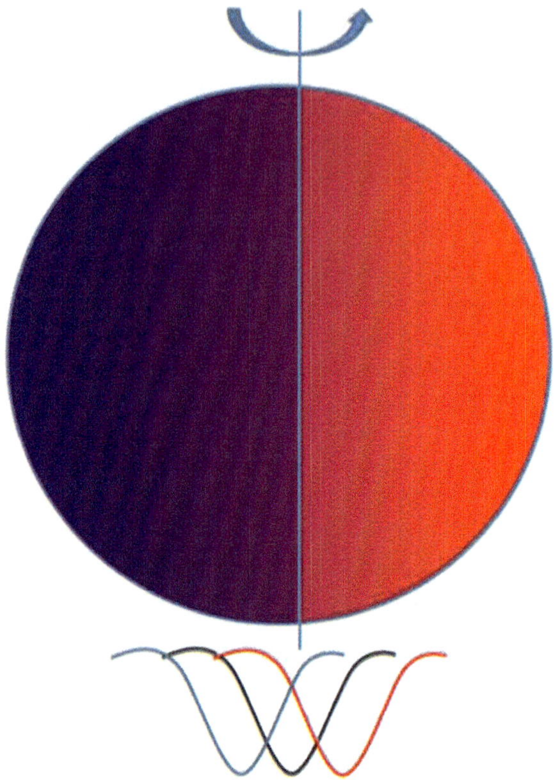

Fig. 4.16 The left part of the solar disk is moving towards us and the spectral lines are blue shifted, while the right part is moving away from us and the spectral lines are red shifted. This is indicated by the blue and red line profiles, respectively. The non-shifted spectral line profile is denoted with black color. Adapted from Hanslmeier (2013)

- influence of meridional flows;
- incorrect treatment of the solar Limb Shift (LS);
- neglect of the variation of the line-asymmetry (C-shape) across the solar disk;
- neglect of the difference of the C-shape between the quiet and active regions;
- scattered light due to the instrument and the Earth's atmosphere;
- possible error in the orientation of the solar rotation axis;
- interference fringes in the spectrograph, instrumental polarization, and other instrumental effects;
- possible errors due to the small errors in published line wavelengths.

One of the most critical issues is the proper treatment of the solar limb shift, which is a complex and still not fully resolved phenomenon. However, it is generally considered that the limb shift of solar spectral lines is caused by granular convective motions. More details can be found in Beckers and Nelson (1978) and Ulrich et al. (1988), while various mathematical representations of the limb shift function were summarized in Schroeter (1985).

4.3.1 The Doppler Effect

The Doppler effect is a shift of the measured wavelength of the spectral line from the object which is moving towards or away from the observer. The measured wavelength λ' is given by the well-known Doppler formula (see standard textbooks, e.g., Hanslmeier (2018), Fox (2018)):

$$\boxed{\lambda' = \lambda_0(1 + v/c)\ ,} \tag{4.5}$$

where λ_0 is the wavelength of the spectral line when the source is at rest relative to an observer (the so-called laboratory wavelength), v is the radial velocity of the source, with the positive sign when it is moving away from the observer and with the negative sign when it is moving towards, and c is the speed of light. The spectral line is red shifted when the source is moving away and blue shifted when the source is moving towards the observer. We described the non-relativistic Doppler effect which is valid when $v \ll c$, which is sufficient for the solar rotation analysis.

4.3.2 Plasma Motions Measured with the Doppler Method

The basic principle of spectroscopic measurement of the solar rotation is to use the least-square fit of the Doppler velocity as a function of central meridian distance for a constant latitude. The procedure is then repeated for various latitudes and the solar differential rotation is obtained as the angular velocity as a function of the latitude.

Although the procedure is rather general, taking into account various geometrical and physical considerations, in practice somewhat different approaches are used, specific for, e.g., the type of the instrument and the measurement method. We recommend three typical examples, Howard and Harvey (1970), Ulrich et al. (1988), and Jejčič and Čadež (2005), see also references therein, and will describe here the procedure for two of them.

The first example is the spectroscopic measurements of the solar rotation using data obtained with the synoptic program of solar observations using the 150-foot tower at Mount Wilson described in detail in Ulrich et al. (1988). From the Dopplergrams corrected for the limb shift a fitted velocity field V_0 is derived:

$$\begin{aligned} V_0 = {} & (A + B(\sin^2 b\ +\ 1.0216295 \sin^4 b) + D \sin^{10}\rho) R_0 \cos b \cos B_0 \sin CMD + \\ & + E + F \sin b + (G \sin b + H \sin 2b)(\cos CMD \sin b \cos B_0 - \cos b \sin B_0)\ . \end{aligned} \tag{4.6}$$

The quantities in this equation have the following meaning:

- V_0—the measured large-scale velocity field;
- A—the equatorial rotation rate;

- B—the parameter B of the solar differential rotation;
- b—the heliographic latitude;
- C—the parameter C of the solar differential rotation, which does not appear here, since it is merged with the parameter B;
- D—the effect of the stray light;
- ρ—the center to limb angular distance;
- R_0—the solar radius;
- B_0—the heliographic latitude of the solar disk center;
- CMD—the central meridian distance;
- E—the overall zero-point;
- F—the spectrograph drift during each scan;
- G—the meridional motions;
- H—the meridional motions.

The final output of the measurements is the solar differential rotation parameters A and B. Note that the parameter C is combined with the parameter B and has almost the same value as B.

In the second example, we describe the procedure of spectroscopic solar rotation measurements using the coelostat and double monochromator DFS-12 of the Ljubljana Observatory Jejčič and Čadež (2005). The radial velocity V for various positions on the solar disk was fitted to the following rotation model:

$$V = V_{\mathrm{gr}} + (A + B\sin^2 b + C\sin^4 b)\, R_0 \sin CMD + e\sin^2\xi + f\sin^4\xi \ . \quad (4.7)$$

The quantities in this equation have the following meaning:

- V—the measured radial velocity;
- V_{gr}—the gravitational redshift velocity;
- A—the equatorial rotation rate;
- B—the parameter B of the solar differential rotation;
- b—the heliographic latitude;
- C—the parameter C of the solar differential rotation;
- R_0—the solar radius;
- CMD—the central meridian distance;
- e—the model parameter related to the limb shift;
- ξ—the distance from the sub-Earth point;
- f—the model parameter related to the limb shift.

The gravitational redshift velocity can be calculated from the gravitational redshift of the wavelength using the general theory of relativity, see, e.g., Hanslmeier (2023), Lang (1999):

$$\boxed{\frac{\Delta\lambda}{\lambda_0} \approx \frac{GM}{R_0 c^2}}\ , \quad (4.8)$$

where $\Delta\lambda$ is the wavelength redshift, λ_0 is the wavelength emitted by the atom when the gravitational field is negligible, G is the gravitational constant, c is the speed of light, and M and R_0 are the mass and the radius of the Sun, respectively.

4.4 Helioseismology

Helioseismology is based on the measurements of the frequencies of various standing wave modes at the solar surface. It is one of the two methods which provides data about the structure of the Sun below its surface. The second method which enables direct information from the solar interior is the measurement of the solar neutrino flux.

By introducing the topic of helioseismology we should make two notes. First, the helioseismology has several applications in solar physics, the determination of solar rotation (below the surface) being only one of them. Examples of other applications embrace investigation of the structure of the Sun, in particular the depth and the composition of the convection zone, temperature, plasma motions, and chemical composition of the solar interior, etc. The second note is that in the last three decades this research field accomplished a huge and rather dynamic development, leading to a series of important discoveries. So, the literature about helioseismology is immense and we will try to mention only the most important research papers and reviews. A comprehensive treatment of the topic is given in Christensen-Dalsgaard (2014), Howe (2009), and Basu (2016). Both of these facts are reflected also in the development of several subtopics within the main field of helioseismology, such as global and local helioseismology, time–distance helioseismology, and ring-diagram analysis. Finally, an extension of the helioseismology to other stars led to the successful development of the asteroseismology, see, e.g., Baade (1992), Christensen-Dalsgaard (2014), and Brun et al. (2015).

4.4.1 Solar Oscillations

Oscillations at the solar surface and in the lower layers of the solar atmosphere are observed as Doppler shifts of spectral lines, which indicates upside-down motions, or as interconnected fluctuations of the radiation intensities. First indication about solar oscillations was the discovery of the 5-minute oscillations in the beginning of 1960es. The lifetime of these oscillations is about 25 minutes, the velocity amplitudes are of the order 0.5 km/s, and the sizes of the oscillation regions are about 10,000 km. In the 1970s, it was found that all arbitrary combinations of oscillation region sizes and oscillation periods are not possible. In the diagram showing frequency vs. wavenumber, the measured values are grouped in a series of branches and the empty regions in this diagram represent combinations which cannot be observed; see Fig. 4.17. A general agreement between the calculated and measured values in this type

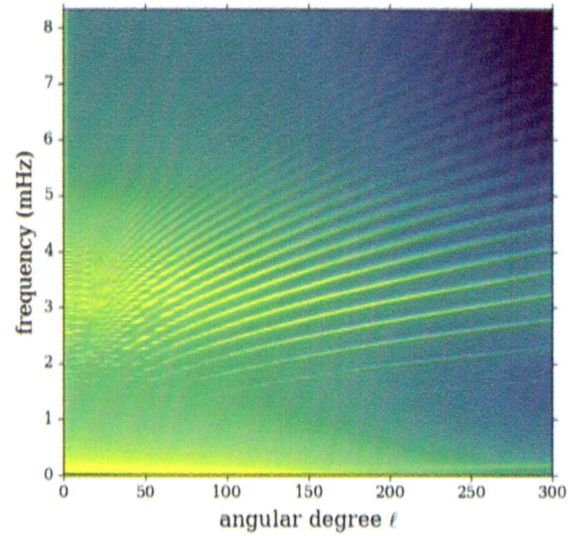

Fig. 4.17 Power spectrum of medium angular degree solar oscillations, computed for 144 days of data from the MDI instrument aboard the satellite SOHO. The color scale is logarithmic and saturated at one hundredth the maximum power in the signal, to make the modes more visible. The low-frequency region is dominated by the signal of granulation. As the angular degree increases, the individual mode frequencies converge onto clear ridges, each corresponding to a sequence of low-order modes.Credit: Wikipedia

of diagram enabled an understanding of the 5-minute solar oscillations as standing acoustic waves. Besides these standing acoustic waves (the so-called p-modes or pressure modes) additionally two further types of oscillation are possible in the Sun. The full list of oscillation modes on the Sun embrace: (Roth (2004), Hanslmeier (2018)):

- p-modes or pressure modes, also acoustic waves: The restoring force is pressure. These waves are generated by the turbulence in the solar convection zone. The interference between the localized waves produces global standing waves which are known as normal modes. Propagation of the acoustic waves in the solar interior depends on the sound speed, which increases with temperature and the temperature increases with depth below the solar surface.
- g-modes or standing internal gravity waves: The restoring force is gravity and negative buoyancy plays an important role. These waves are confined below the convection zone and have very small residual amplitudes in the photosphere, of only a few millimeters. They are extremely hard to detect.
- f-modes or surface gravity waves: These are gravity waves at or near the solar photosphere. They can be visualized as the water waves at the surface of a deep ocean.

Two-dimensional oscillating structures observed on the solar surface are caused by the three-dimensional standing wave patterns in the solar interior which acts as a resonant cavity; see Fig. 4.18. Standing acoustic waves (the p-modes) are trapped in the solar convection zone. Convection zone of the Sun is the subsurface layer in which the energy is transported outwards with plasma motions caused by temperature and density gradients in the gravitational field. Solar interior does not have solid walls from which the waves could be reflected. However, the acoustic waves are reflected

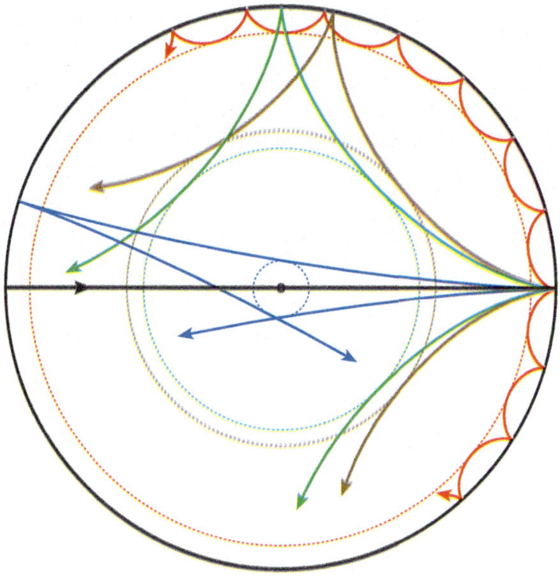

Fig. 4.18 An illustration of the solar oscillations and propagation of waves in the interior of the Sun. For larger oscillating areas the waves spread deeper into the solar interior. Adapted from https://solar-center.stanford.edu/helio-ed-mirror/english/engrays.html.

from the solar surface from below due to the large density gradient and are reflected deep in the convective zone due to the increased speed of sound. In the mentioned diagram frequency vs. wavenumber small but systematic differences between the calculated and measured values were found, which was interpreted with the larger depth of the convection zone than previously considered. The depth of the solar convection zone below the surface amounts to about one-third of the solar radius.

4.4.2 Solar Rotation Measured with the Helioseismological Method

The diagram frequency vs. wavenumber which is observed on the Sun is typical for the so-called non-radial pulsations. In these pulsations, various parts of the Sun, at the same radial distance from the solar center, are moving (oscillating) in different ways. So, the average solar radius remains practically unchanged. On the other hand, many variable stars exhibit the so-called radial oscillations, or pulsations, in which case all parts of the star at the same radial distance from the center are moving in the same way, synchronously. We are speaking about the pulsating variable stars, which periodically change their radius, luminosity, and surface temperature on the temporal scale of days.

The velocity pattern of a particular non-radial oscillation mode on the Sun can be, mathematically speaking, separated into the radial and angular parts. Each mode is described by the eigenvalues n, l, and m. The radial part is defined with the number n, and the angular part with the numbers l and m; see Fig. 4.19. The oscillation pattern at the solar surface is characterized by the node lines at which the material is not moving. The total number of the node lines at the surface is denoted by harmonic number (or degree) l. The azimuthal number (degree) m gives the number of node lines which are crossing the solar pole. Note that it always holds $|m| \leq l$. Finally, the radial number (order) n denotes the number of node lines in the radial direction in the solar interior. Further details about the eigenvalues n, l, and m of solar oscillations can be found in Baade (1992), Stix (2000), Roth (2004), and Hanslmeier (2023).

Perhaps the reader has already noticed that we speak about the frequency–wavenumber diagram, but in the Fig. 4.17 the angular degree (number l) is given on the x-axis. This will be clarified now. The dependence of the colatitude θ and longitude ϕ for a single mode is given by a spherical harmonic $Y_l^m(\theta, \phi)$. Now we should see how the wavenumber and the degree l are related. The horizontal wavelength λ_h and the magnitude of the horizontal component of the wavenumber k_h at the distance r from the solar center can be expressed as

$$\boxed{k_\mathrm{h} = \frac{2\pi}{\lambda_\mathrm{h}} = \frac{L}{r}}\,, \tag{4.9}$$

where $L = \sqrt{l(l+1)}$. In this way, a relationship between the wavenumber and the harmonic (angular) degree number l is established. More details about this topic can be found in Christensen-Dalsgaard (1992), Mattig (1995), and Christensen-Dalsgaard (1996).

Fig. 4.19 Illustration of a solar pressure mode (p mode) with the radial order $n = 14$, angular degree $l = 20$, and azimuthal order $m = 16$. Credit: Wikipedia

Finally, we come to the most important point regarding the solar internal rotation determination Christensen-Dalsgaard (1990), Hanasoge et al. (2015), Basu et al. (2015), Gough (2015). If the Sun would not rotate, the modes with the same numbers n and l, but with a different number m, would exhibit the same frequency. However, since the Sun does rotate, the solar rotation removes this degeneracy, and the measured small differences in the frequency of the modes with different numbers m reveal the rotation velocity in the solar interior. In other words, we can describe this as follows. In a non-rotating Sun, the frequencies would be independent of the azimuthal order m. The solar rotation introduces, however, a splitting of the frequencies according to m, not unlike the Zeeman splitting of the energy levels of an atom in a magnetic field Christensen-Dalsgaard (1992). For energy-splitting and removing of degeneracy in quantum and atomic physics, see, e.g., Schiff (1968), Weinberg (2012), and Fox (2018).

The techniques for observing solar oscillations were described in Brown (1996). In testing the solar model with the helioseismology we distinguish between the so-called forward Christensen-Dalsgaard (1996) and inverse analysis Gough (1996). In practice, the frequency splitting $\Delta\nu_{nlm}$ is measured Beck (2000). This frequency splitting is the difference between modes of the same multiplet:

$$\Delta\nu_{nlm} = \nu_{nl(+m)} - \nu_{nl(-m)} \ . \tag{4.10}$$

The splittings are related to the underlying profile of the differential rotation by the equation:

$$\Delta\nu_{nlm} = \int\int K_{nlm}(r, b)\, \omega(r, b)\, r\, \mathrm{d}r\, \mathrm{d}b \ , \tag{4.11}$$

where r is the radial distance, b is the latitude, $\omega(r, b)$ is the rotation velocity at the position defined by r and b, and the functions $K_{nlm}(r, b)$ are the mode kernels Gough (1996) and Burston et al. (2015). These mode kernels are generally known quantities when a solar model is defined. The procedure of inversion means to determine the differential rotation profile $\omega(r, b)$ from the measurements of frequency splitting $\Delta\nu_{nlm}$. Several numerical procedures are possible, such as regularized least-square fit (RLS) and optimally localized averages (OLA).

4.4.3 A Historical Remark

The 5-minute oscillations on the Sun were discovered in the early 1960s by Robert B. Leighton and coworkers, while the foundations of helioseismology are based on the works of Franz-Ludwig Deubner, Roger Ulrich, and other astronomers in the 1970s. Detailed description of these important historical efforts, including both the observational and theoretical aspects, can be found in various reviews, such as

Leibacher and Stein (1981), Stein and Leibacher (1981), Deubner (1981), Reiter (2004), and Christensen-Dalsgaard et al. (1985).

4.5 Correlation of Solar Images

Consider a time series of full-disk solar images, for instance filtergrams taken in the light of a spectral line or magnetograms. In the tracer method, described earlier, the observer follows discernible objects on the Sun, such as sunspots, Hα filaments, and coronal bright points. On the other hand, using the method of correlating solar images, no specific solar features are followed, but all structures in the images are correlated and identified again in the subsequent full-disk images. It often turns out that the solar structures identified in a series of images correspond to the small-scale or large-scale magnetic elements, or features related to them.

Since the measurements are not focused on only one type of solar object, it might happen that various differential rotation profiles can coexist simultaneously, corresponding to different structures identified in a series of images. This is described by the well-known dispersion relation and an interpretation might be rather complicated and deserves a special attention.

As in the description of earlier methods, we mention briefly some advantages and disadvantages of the correlation method for solar rotation determination. The main advantage is that the method is generally not influenced by a butterfly diagram. This means that almost all solar latitudes and solar cycle phases are well covered in the analysis, which is rarely the case in using most tracers. Of course, the weather conditions and technical issues might cause some gaps in observations but this is unavoidable and common to all methods. The most important disadvantage of the correlation method is that in some cases we do not have a visual identification of the correlated features. This might lead to misidentification of structures and to artificial velocity signals, which are of non-solar origin.

4.5.1 Cross-correlation and Autocorrelation

In modern observational astronomy and astrophysics, the cross-correlation and autocorrelation techniques are widely used in different contexts Ivezić et al. (2014). Here we will focus on applications of the methods for the solar rotation measurements and will give the appropriate definitions and examples in this context; see Snodgrass (1992) and Stenflo (1989).

In the cross-correlation method, a time series of the full-disk solar images is correlated on the time scale of several days. The method is not unlike the simple passage method, discussed earlier when the tracer method was introduced. A good example of the method is the cross-correlation of the Mount Wilson full-disk solar magnetograms made over successive days Snodgrass (1983). One of the main advantages of the method is a good coverage of high solar latitudes. Further examples of the successful application of the cross-correlation method include the solar rotation measurements from a series of OSO-4 EUV spectroheliograms Dupree and Henze (1972) and solar rotation measurements of the small magnetic features from the Kitt Peak high resolution magnetograms Komm et al. (1993).

On the other hand, the autocorrelation method is based on an identification of the structures which live long enough to reappear at the central meridian after at least one solar rotation. This approach is in a way similar to the period method discussed earlier. Classical examples for the autocorrelation method are the correlations of the successive synoptic maps of solar magnetic field; see Sheeley et al. (1987) and Stenflo (1989). As the magnetic field pattern recurs at the central meridian after each solar rotation, peaks in the autocorrelation functions are obtained at the lags corresponding to an integer number of the solar rotation periods.

4.5.2 Synoptic Charts

Synoptic charts represent solar structures, or a layer in the solar atmosphere, unfolded in a two-dimensional coordinate system. The solar longitude is on the x-axis, while the solar latitude is on the y-axis. The longitudes span 0–360 degrees and the latitudes 0–90 degrees, both on the northern and on the southern solar hemisphere. Typically, synoptic charts are produced for each Carrington rotation. Also, synoptic charts are often constructed for the solar magnetic field on various spatial scales.

An example of a time series of synoptic charts representing the large-scale solar magnetic field is given in Fig. 4.20. This figure represents a part of a solar rotation analysis based on a comparison of a series of magnetic synoptic charts Brajsa et al. (1992). Note that the coordinate transformation is performed using the Carrington system, as described earlier in this book. This means that the uniform rotation velocity is assumed for all latitudes. So, if the differential rotation would not be present on the Sun, the identified structures would be on the same place (longitude) in successive synoptic charts. However, since the Sun does show a differential rotation, we note systematic displacements of the recognized features from one synoptic chart to the other, as can clearly be seen in Fig. 4.20. This is a direct evidence of a faster or slower rotation velocity of these features. Moreover, it is interesting to note that the solar rotation also shows a temporal variation.

Finally, we note that also fixed coordinate systems other than Carrington can be used in synoptic chart composition. In that case the velocity deviations are referred to this other uniform rotation velocity.

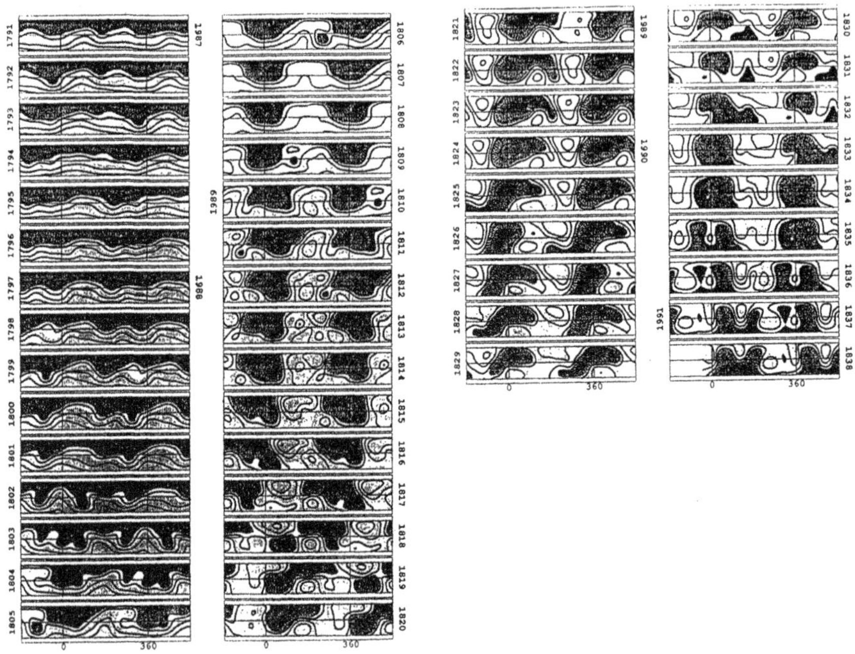

Fig. 4.20 Series of synoptic magnetic maps in which the areas of opposite magnetic polarities can be distinguished. The time is progressing from up to down. The numbers of the Carrington rotations are given on the left-hand side and the calendar years on the right-hand side of each column. The original data were taken from John M. Wilcox Solar Observatory at Stanford, USA Hoeksema (1991). Adapted from Brajsa et al. (1992)

4.5.3 Stackplots

The so-called stackplot method is a special subclass of methods which combines the synoptic charts method and the autocorrelation method. Stackplots are constructed by slicing the Carrington synoptic charts into latitude strips and by reassembling these strips into vertical columns. This enables a picture of magnetic field (or some other feature revealed by a filtergram) at each latitude as a function of the Carrington longitude and the number of the Carrington rotation. In this way, a temporal variation during the solar cycle can also be revealed. This method for a new determination of the solar rotation velocity, using magnetic field measurements and including some specific modifications of the method, is described in detail in Sheeley et al. (1992) and references therein.

The stackplot method was successfully used for the solar rotation determination using Hα, magnetic field, and He I 1083 nm synoptic charts McIntosh (1992) and by using the Yohkoh soft X-ray images Kozuka et al. (1994) and Weber et al. (1999).

Exercises and Questions

1. On the Sun a total of 45 sunspots in 7 sunspot groups were observed. Calculate the relative sunspot number using Eq. (4.1), assuming that the calibration factor is equal to 1.

2. Compare the advantages and disadvantages of the tracer method with those for the Doppler method.

3. If you would have at your disposal all astronomical facilities in the world, both terrestrial and from the satellites, which method/telescope would you use for the most precise determination of the solar rotation?

4. Does the amount of the gravitational redshift depend on the Doppler measurement position on the solar disk?

5. We have seen that the standing acoustic waves, the so-called p-modes, are trapped in the solar convection zone. How and where these waves are reflected in the solar interior?

6. What is the difference between the cross-correlation and autocorrelation methods in the context of the solar rotation determination.

7. Explain how the systematic displacement of large-scale structures in a series of successive Carrington synoptic maps indicate the solar differential rotation.

References

Arlt, R., Senthamizh Pavai, V., Schmiel, C., Spada, F.: Sunspot positions, areas, and group tilt angles for 1611–1631 from observations by Christoph Scheiner. Astron. Astrophys. **595**, A104 (2016)

Baade, D.: Observational aspects of stellar seismology. Rev. Mod. Astron. **5**, 125–142 (1992)

Balthasar, H., Woehl, H.: Differential rotation and meridional motions of sunspots in the years 1940–1968. Astron. Astrophys. **92**(1–2), 111–116 (1980)

Balthasar, H., Schuessler, M., Woehl, H.: On changes of the rotation velocities of stable, recurrent sunspots and their interpretation with a flux tube model. Solar Phys. **76**(1), 21–28 (1982)

Balthasar, H., Vazquez, M., Woehl, H.: Differential rotation of sunspot groups in the period from 1874 through 1976 and changes of the rotation velocity within the solar cycle. Astron. Astrophys. **155**(1), 87–98 (1986)

Basu, S.: Global seismology of the Sun. Living Rev. Solar Phys. **13**(1), 2 (2016)

Basu, S., Grevesse, N., Mathis, S., Turck-Chièze, S.: Understanding the internal chemical composition and physical processes of the solar interior. Space Sci. Rev. **196**(1–4), 49–77 (2015)

Beck, J.G.: A comparison of differential rotation measurements-(Invited Review). Solar Phys. **191**(1), 47–70 (2000)

Beckers, J.M., Nelson, G.D.: Some comments on the limb shift of solar lines. II: The effect of granular motions. Solar Phys. **58**(2), 243–261 (1978)

Brajša, R., Wöhl, H., Vršnak, B., Ruždjak, V., Clette, F., Hochedez, J.F.: Solar differential rotation determined by tracing coronal bright points in SOHO-EIT images. II. Results for 1998/99 obtained with interactive and automatic methods. Astron. Astrophys. **392**, 329–334 (2002)

Brajša, R., Wöhl, H., Vršnak, B., Ruždjak, V., Clette, F., Hochedez, J.F.: Solar differential rotation determined by tracing coronal bright points in SOHO-EIT images. I. Interactive and automatic methods of data reduction. Astron. Astrophys. **374**, 309–315 (2001)

Brajša, R., Vršnak, B., Ruždjak, V., Schroll, A., Pohjolainen, S., Urpo, S., Teräsranta, H.: Solar differential rotation determined by polar crown filaments. Solar Phys. **133**(2), 195–203 (1991)

Brajsa, R., Vrsnak, B., Ruzdjak, V., Bozic, H., Pohjolainen, S., Urpo, S., Terasranta, H.: Variations of solar global rotation during the polarity reversal. Hvar Obs. Bull. **16**(1), 13 (1992)

Brajša, R., Ruždjak, V., Vršnak, B., Wöhl, H., Pohjolainen, S., Upro, S.: Statistical weights and selective height Corrections in the determination of the solar rotation velocity. Solar Phys. **196**(2), 279–297 (2000)

Brajša, R., Wöhl, H., Vršnak, B., Ruždjak, D., Sudar, D., Roša, D., Hržina, D.: Differential rotation of stable recurrent sunspot groups. Solar Phys. **206**(2), 229–241 (2002)

Brajša, R., Wöhl, H., Vršnak, B., Rušdjak, V., Clette, F., Hochedez, J.F., Verbanac, G., Temmer, M.: Spatial distribution and north-south asymmetry of coronal bright points from mid-1998 to mid-1999. Solar Phys. **231**(1–2), 29–44 (2005)

Brajša, R., Ruždjak, D., Wöhl, H.: Temporal variations of the solar rotation determined by sunspot groups. Solar Phys. **237**(2), 365–382 (2006)

Brajša, R., Wöhl, H., Hanslmeier, A., Verbanac, G., Ruždjak, D., Cliver, E., Svalgaard, L., Roth, M.: On solar cycle predictions and reconstructions. Astron. Astrophys. **496**(3), 855–861 (2009)

Brajša, R., Romštajn, I., Wöhl, H., Benz, A.O., Temmer, M., Roša, D.: Heights of solar tracers observed at 8 mm and an interpretation of their radiation. Astron. Astrophys. **493**(2), 613–621 (2009)

Brajša, R., Skokić, I., Sudar, D., Benz, A.O., Krucker, S., Ludwig, H.G., Saar, S.H., Selhorst, C.L.: ALMA small-scale features in the quiet Sun and active regions. Astron. Astrophys. **651**, A6 (2021)

Brajša, R., Verbanac, G., Bandić, M., Hanslmeier, A., Skokić, I., Sudar, D.: A prediction for the 25th solar cycle maximum amplitude. Astronomische Nachrichten **343**(3), e13960 (2022)

Brown, T.M.: Techniques for observing solar oscillations. In: The Structure of the Sun, 6th Canary Islands Winter School of Astrophysics, Cambridge University Press, p. 1 (1996)

Brun, A.S., García, R.A., Houdek, G., Nandy, D., Pinsonneault, M.: The solar-stellar connection. Space Sci. Rev. **196**(1–4), 303–356 (2015)

Bruzek, A.: Über die Umlaufgeschwindigkeit der polaren Filamente. Mit 2 Textabbildungen. Zeitschrift f. Astrophys. **51**, 75 (1961)

Bruzek, A., Durrant, C.J.: Illustrated glossary for solar and solar-terrestrial physics. D. Reidel Publ. Comp, Dordrecht (1977)

Burston, R., Gizon, L., Birch, A.C.: Interpretation of helioseismic travel times. Sensitivity to sound speed, pressure, density, and flows. Space Sci. Rev. **196**(1-4), 201–219 (2015)

Christensen-Dalsgaard, J.: Lecture notes on stellar oscillations, 5th ed. (2014)

Christensen-Dalsgaard, J.: Seismic investigation of the solar interior. In: Schmelz, J.T., Brown, J.C. (eds.) *The Sun: A Laboratory for Astrophysics*, vol. 373 of *NATO Advanced Study Institute (ASI) Series C*, p. 29 (1992)

Christensen-Dalsgaard, J.: Testing a solar model: the forward problem. In: The Structure of the Sun, 6th Canary Islands Winter School of Astrophysics, Cambridge University Press, p. 47 (1996)

Christensen-Dalsgaard, J.: Helioseismic measurements of the solar internal rotation. Rev. Mod. Astron. **3**, 313–349 (1990)

Christensen-Dalsgaard, J., Gough, D., Toomre, J.: Seismology of the Sun. Science **229**(4717), 923–931 (1985)

Clette, F., Leif, S., José, M.V., Edward, W.C.: Revisiting the sunspot number. A 400-year perspective on the solar cycle. Space Sci. Rev. **186**(1–4), 35–103 (2014)

Cliver, E.W.: The extended cycle of solar activity and the sun's 22-year magnetic cycle. Space Sci. Rev. **186**(1–4), 169–189 (2014)

Cliver, E.W., Clette, F., Svalgaard, L.: Recalibrating the sunspot number (SSN): The SSN workshops. Central Eur. Astrophys. Bull. **37**, 401–416 (2013)

Cliver, E.W., Clette, F., Svalgaard, L., Vaquero, J.M.: Recalibrating the sunspot number (SN): The 3rd and 4th SN workshops. Central Eur. Astrophys. Bull. **39**, 1–19 (2015)
Deubner, F.L.: Pulsations and oscillations. In: Stuart, J. (ed.) The Sun as Star, NASA Special Publication **450**, 65–84 (1981)
Dezso, L., Kovacs, A.: On the differential Rotation of the solar photosphere. Solar Phys. **151**(2), 385–388 (1994)
Dumbović, M., Karbonini, L., Čalogović, J., Matković, F., Martinić, K., Remeshan, A.K., Brajša, R., Vršnak, B.: The catalog of Hvar observatory solar observations, in press. Solar Phys. (2024)
Dupree, A.K., Jr., Henze, W.: Solar rotation as determined from OSO-4 EUV spectroheliograms. Solar Phys. **27**(2), 271–279 (1972)
Ermolli, I., Shibasaki, K., Tlatov, A., van Driel-Gesztelyi, L.: Solar cycle indices from the photosphere to the corona: Measurements and underlying physics. Space Sci. Rev. **186**(1–4), 105–135 (2014)
Foukal, P.: Solar astrophysics, 3rd revised edn. Wiley-VCH-Varlag (2013)
Fox, M.: A student's guide to atomic physics. Cambridge University Press, Cambridge (2018)
Gibson, S.E.: Solar prominences: Theory and models. Fleshing out the magnetic skeleton. Living Rev. Solar Phys. **15**(1), 7 (2018)
Godoli, G., Mazzucconi, F., Piergianni, I.: Sunspot rotation rates along the butterfly diagram. Solar Phys. **148**(2), 195–200 (1993)
Golub, L., Pasachoff, J.M.: The solar corona, 2nd edn. Cambridge University Press, Cambridge (2009)
Golub, L., Krieger, A.S., Silk, J.K., Timothy, A.F., Vaiana, G.S.: Solar X-Ray Bright Points. Astrophys. J. Lett. **189**, L93 (1974)
Gough, D.: Testing solar models: The inverse problem. In: The Structure of the Sun, 6th Canary Islands Winter School of Astrophysics, Cambridge University Press, p. 141 (1996)
Gough, D.O.: Some glimpses from helioseismology at the dynamics of the deep solar Interior. Space Sci. Rev. **196**(1–4), 15–47 (2015)
Gurman, J.B.: Sunspots: A Laboratory for solar physics. In: Schmelz, J.T., Brown, J.C. (eds.) *The Sun: A Laboratory for Astrophysics*, vol. 373 of *NATO Advanced Study Institute (ASI) Series C*, pp. 221 (1992)
Hanasoge, S., Miesch, M.S., Roth, M., Schou, J., Schüssler, M., Thompson, M.J.: Solar dynamics, rotation, convection and overshoot. Space Sci. Rev. **196**(1–4), 79–99 (2015)
Hanslmeier, A.: Die Sonne-der Stern von dem wir leben. Vehling-Verlag, Graz (2013)
Hanslmeier, A.: Planetary habitability and stellar activity, vol. 3. World Scientific Publishing, Singapore, Advances in Planetary Science (2018)
Hanslmeier, A.: The chaotic solar cycle. Springer Nature, Singapore (2020)
Hanslmeier, A.: Introduction to astronomy and astrophysics. Springer-Varlag, Berlin (2023)
Harvey, K.L.: The cyclic behavior of solar activity. In: Harvey, K.L. (ed.) *The Solar Cycle*, vol. 27 of *Astronomical Society of the Pacific Conference Series*, p. 335 (1992)
Hoeksema, J.T.: The solar magnetic field 1985 through 1990, center for space sciences and astrophysics report CSSA-ASTRO-91-01. J. M. Wilcox Solar Obs, Stanford (1991)
Howard, R.: Solar rotation. Ann. Rev. Astron. Astrophys. **22**, 131–155 (1984)
Howard, R.F.: The magnetic fields of active regions-part two. Solar Phys. **126**(2), 299–309 (1990)
Howard, R., Harvey, J.: Spectroscopic determinations of solar rotation. Solar Phys. **12**(1), 23–51 (1970)
Howard, R., Gilman, P.I., Gilman, P.A.: Rotation of the sun measured from Mount Wilson white-light images. Astrophys. J. **283**, 373–384 (1984)
Howard, R.F., Harvey, J.W., Forgach, S.: Solar surface velocity fields determined from small magnetic features. Solar Phys. **130**(1–2), 295–311 (1990)
Howe, R.: Solar interior rotation and its variation. Living Rev. Solar Phys. **6**(1), 1 (2009)
Husak, M.,Brajša, R., Špoljarić, D., Krajnović, D., Ruždjak, D., Skokić, I., Roša, D., Hržina, D.: Bošković's method for determining the axis and rate of solar rotation by observing sunspots in 1777. Solar Phys. textbf298(10), 122 (2023)

Ivezić, Ž, Connolly, A.J., VanderPlas, J.T., Gray, A.: Statistics, data mining, and machine learning in astronomy: A practical python guide for the analysis of survey data. Princeton Univ. Press, Princeton and Oxford (2014)
Javaraiah, J., Gokhale, M.H.: Periodicities in the north-south asymmetry of the solar differential rotation and surface magnetic field. Solar Phys. **170**(2), 389–410 (1997)
Kambry, M.A., Nishikawa, J.: Solar differential rotation derived from sunspot observations. Solar Phys. **126**, 89–100 (1990)
Jejčič, S., Čadež, A.: The comparison of spectroscopic measurements of the solar rotation. Hvar Obs. Bull. **29**, 11–20 (2005)
Komm, R.W., Howard, R.F., Harvey, J.W.: Rotation rates of small magnetic features from two-dimensional and one-dimensional cross-correlation analyses. Solar Phys. **145**(1), 1–10 (1993)
Kozuka, Y.T., Watanabe, M.K., Ohyama, M., Saito, T.: Rotation rates of soft X-ray coronal structures. In: *Proceedings of Kofu Symposium*, pp. 359–362 (1994)
Lang, K.R.: Astrophysical formulae. Springer-Verlag, Berlin (1999)
Leibacher, J.W., Stein, R.F.: Oscillations and pulsations. In: Stuart, J. (ed.) The Sun as a Star, NASA Special Publication **450**, 263–287 (1981)
Liu, S.Y., Kundu, M.R.: Differential rotation of the solar atmosphere as determined from millimeter data. Solar Phys. **46**(1), 15–22 (1976)
Madjarska, M.S.: Coronal bright points. Living Rev. Solar Phys. **16**(1), 2 (2019)
Matković, F., Brajša, R., Temmer, M., Heinemann, S.G., Ludwig, H.G., Saar, S.H., Selhorst, C.L., Skokić, I., Sudar, D.: Differences in physical properties of coronal bright points and their ALMA counterparts within and outside coronal holes. Astron. Astrophys. **670**, A146 (2023)
Mattig, W.: Die Sonne. Verlag C. H, Beck, München (1995)
McIntosh, P.S.: Solar interior processes suggested by large-scale surface patterns. In: Harvey, K.L. (ed.) *The Solar Cycle*, volume 27 of *Astronomical Society of the Pacific Conference Series*, p. 14 (1992)
Meunier, N., Nesme-Ribes, E., Grosso, N.: Cycle 19 facula dynamics. I. Angular rotation. Astron. Astrophys. **319**, 673–682 (1997)
Parenti, S.: Solar prominences: Observations. Living Rev. Solar Phys. **11**(1), 1 (2014)
Pohjolainen, S., Urpo, S., Terasranta, H., Vrsnak, B., Brajsa, R., Ruzdjak, V., Jurac, S.: Intensity variations and short time evolution of solar microwave low temperature regions. Hvar Obs. Bull. **15**(1), 21 (1991)
Poljancic Beljan, I., Jurdana-Šepić, R., Brajša, R., Sudar, D., Ruždjak, D., Hržina, D., Pötzi, W., Hanslmeier, A., Veronig, A., Skokić, I., Wöhl, H.: Solar differential rotation in the period 1964–2016 determined by the Kanzelhöhe data set. Astron. Astrophys. **606**, A72 (2017)
Poljančić Beljan, I., Jurdana-Šepić, R., Jurkić, T., Brajša, R., Skokić, I., Sudar, D., Ruždjak, D., Hržina, D., Pötzi, W., Hanslmeier, A., Veronig, A.M.: Variation in solar differential rotation and activity in the period 1964–2016 determined by the Kanzelhöhe data set. Astron. Astrophys. **663**, A24 (2022)
Reiter, J.: Die Anfänge der Helioseismologie. Sterne und Weltraum **43**(8), 33–37 (2004)
Ruzdjak, D., Brajsa, R., Sudar, D., Skokic, I., Poljancic Beljan, I.: A relationship between the solar rotation and activity analysed by tracing sunspot groups. Solar Phys. **292**(12), 179 (2017)
Ruzdjak, D., Sudar, D., Brajsa, R., Skokic, I., Poljancic Beljan, I., Jurdana-Sepic, R., Hanslmeier, A., Veronig, A., Poetzi, W.: Meridional motions and reynolds stress determined by using kanzelhöhe drawings and white light solar images from 1964 to 2016. Solar Phys. **293**(4), 59 (2018)
Roth, M.: Neue blicke in das innere der sonne. Sterne und Weltraum **43**(8), 24–32 (2004)
Ruzdjak, V., Tandberg-Hanssen, E.: Dynamics of quiescent prominences: IAU Coll. 117, vol. 363. LNP, Springer-Verlag, Berlin (1990)
Ruždjak, D., Brajša, R., Skokić, I., Sudar, D., Hanslmeier, A.: On the periodicities in solar rotation and activity determined from sunspot-group data and possible connection with rossby-like waves. Solar Phys. textbf298(3), 39 (2023)
Ruždjak, D., Ruždjak, V., Brajša, R., Wöhl, H.: Deceleration of the rotational velocities of sunspot groups during their evolution. Solar Phys. **221**(2), 225–236 (2004)

Scheiner, Christoph: Rosa ursina (1630)
Schiff, L.I.: Quantum mechanics. McGraw-Hill Book Comp, Singapore (1968)
Schroeter, E.H.: The solar differential rotation-present status of observations. Solar Phys. **100**, 141 (1985)
Sheeley, N.R., Nash, A.G., Wang, Y.M.: The origin of rigidly rotating magnetic field patterns on the sun. Astrophys. J. **319**, 481 (1987)
Sheeley, N.R., Wang, Y.M., Nash, A.G.: A new determination of the solar rotation rate. Astrophys. J. **401**, 378 (1992)
Sivaraman, K.R., Gupta, S.S., Robert, F.H.: Measurement of kodiakanal white-light images-part one. Solar Phys. **146**(1), 27–47 (1993)
Snodgrass, H.B.: Magnetic rotation of the solar photosphere. Astrophys. J. **270**, 288–299 (1983)
Snodgrass, H.B.: Smokestacks and balloonmen: A magnetic rotation controversy. In: Harvey, K.L. (ed.) *The Solar Cycle*, volume 27 of *Astronomical Society of the Pacific Conference Series*, p. 71 (1992)
Snodgrass, H.B.: Synoptic observations of large scale velocity patterns on the Sun. In: Harvey, K.L. (ed.) *The Solar Cycle*, vol. 27 of *Astronomical Society of the Pacific Conference Series*, p. 205 (1992)
Stein, R.F., Leibacher, J.W.: Wave generation. In: Stuart, J. (ed.) The Sun as a Star, NASA Special Publication **450**, 289–300 (1981)
Stenflo, J.O.: Small-scale magnetic structures on the Sun. Astron. Astrophys. Rev. **1**(1), 3–48 (1989)
Stenflo, J.O.: Differential rotation of the sun's magnetic field pattern. Astron. Astrophys. **210**(1–2), 403–409 (1989)
Stix, M.: The sun's differential rotation. Rev. Mod. Astron. **2**, 248–266 (1989)
Stix, M.: The Sun's normal modes of oscillation. Naturwissenschaften **87**(12), 524–531 (2000)
Stix, M.: The Sun: An introduction. Springer-Varlag, Berlin (2002)
Sudar, D., Brajsa, R., Skokic, I., Poljancic Beljan, I., Woehl, H.: Meridional motion and reynolds stress from debrecen photoheliographic data. Solar Phys. **292**(7), 86 (2017)
Sudar, D., Skokić, I., Brajša, R., Saar, S.H.: Steps towards a high precision solar rotation profile: Results from SDO/AIA coronal bright point data. Astron. Astrophys. **575**, A63 (2015)
Sudar, D., Brajša, R.: Solar rotation in the period 1611–1631 determined using observations of christoph scheiner. Solar Phys. **297**(10), 132 (2022)
Sudar, D., Skokić, I., Ruždjak, D., Brajša, R., Wöhl, H.: Tracing sunspot groups to determine angular momentum transfer on the Sun. Mon. Not. **439**(3), 2377–2385 (2014)
Suzuki, M.: The rOtation of sunspots during activity cycle 22. Solar Phys. **178**(2), 259–265 (1998)
Tandberg-Hanssen, E.: The nature of solar prominences, vol. 199. Kluwer Acad. Publ, Dordrecht, ASSL (1995)
Thomas, J.H., Weiss, N.O.: Sunspots and starspots. Cambridge University Press, Cambridge (2008)
Ulrich, R. K., Boyden, J. E., Webster, L., Snodgrass, H. B., Padilla, S. P., Gilman, P., Shieber, T.: Solar rotation measurements at MT.WILSON–Part five. Solar Phys. **117**(2), 291–328 (1988)
Waldmeier, M.: Ergebnisse und Probleme der Sonnenforschung, 2nd edn. Akademische Verlagsgesellschaft Geest & Portig K.-G, Leipzig (1955)
Weber, M.A., Acton, L.W., Alexander, D., Kubo, S., Hara, H.: A method for characterizing rotation rates in the soft X-Ray corona. Solar Phys. **189**(2), 271–288 (1999)
Weinberg, S.: Lectures on quantum Mechanics. Cambridge University Press, Cambridge (2012)
Wilson, P.R.: Solar and stellar activity cycles. Cambridge Astrophys. Ser. **24** (1994)
Woehl, H.: On the determination of heliographic positions and rotation velocities of sunspots-part one-comparison of results from different observatories and different observing procedures. Solar Phys. **88**(1–2), 65–70 (1983)
Wöhl, H., Brajša, R., Hanslmeier, A., Gissot, S.F.: A precise measurement of the solar differential rotation by tracing small bright coronal structures in SOHO-EIT images. Results and comparisons for the period 1998–2006. Astron. Astrophys. **520**, A29 (2010)
Wöhl, H.: Large-scale motions in the solar photosphere (Observational Aspects). Publ. Deb. Heliophys. Obs. **7**, 19–26 (1990)

Zappala, R.A., Zuccarello, F.: Angular velocities of sunspot-groups and solar photospheric rotation. Astron. Astrophys. **242**(2), 480–487 (1991)
Zirin, H.: Solar activity. In: Schmelz, J.T., Brown, J.C. (eds.) *The Sun: A Laboratory for Astrophysics*, vol. 373 of *NATO Advanced Study Institute (ASI) Series C*, p. 449 (1992)
Zirin, H.: Astrophysics of the Sun. Cambridge University Press, Cambridge (1988)

Chapter 5
Analysis and Reduction of Observational Data

In the previous chapter, we described in detail how the solar rotation velocity can be measured. Now we come to the next step which is the analysis and reduction of observational data which are gathered using the methods described in the previous chapter. In the first part, we describe the transformation from the observed (synodic) to the real (sidereal) solar rotation velocity. The correct synodic–sidereal transformation is needed to avoid this source of systematic error. Then we treat the topic of height correction; if not taken into account properly, using tracers located above (or below) the photosphere could lead to false rotation velocities. After these two sources of errors are removed further data reduction can be performed. This includes an application of the appropriate mathematical representation of the solar differential rotation, the use of the least-square method, removing the cross-talk between the solar differential rotation parameters, and filtering the measured solar rotation velocity data to exclude erroneous outliers.

5.1 Kepler Equation and the Synodic–Sidereal Transformation of the Solar Rotation Velocity

Measurements of the solar rotation from the Earth give the so-called synodic rotation velocity ω_{syn}. This is the apparent solar rotation speed relative to the Earth which revolves around the Sun, making the full period in one year. But, the true solar rotation velocity is relative to the stars and is called the sidereal rotation velocity denoted by ω_{sid}. As we have seen in Chap. 1, an approximative transformation between the two velocities is given by a constant value using a simple formula: $\omega_{\mathrm{sid}} \approx \omega_{\mathrm{syn}} + (360^\circ/365\mathrm{d}) \approx \omega_{\mathrm{syn}} + 0.99^\circ/\mathrm{d}$.

However, the Earth's orbit around the Sun is not a circle but is an ellipse. So, the orbital velocity of the Earth has a small, season's dependent variation, according to the second Kepler law. So, the real synodic–sidereal transformation is not a constant, but varies as a function of time during the year. The correct transformation should

R. Brajša and A. Hanslmeier, *Solar Rotation*, UNITEXT for Physics,
https://doi.org/10.1007/978-981-97-6879-0_5

take into account this variable orbital velocity of the Earth knowing the time of solar observation. During the last couple of decades at least four different procedures for the exact synodic–sidereal transformation were developed and we describe them in detail in the following subsections. The precision of each method as well as possible issues in practical application of them will also be discussed.

5.1.1 Method I: According to Graf (1974)

One of the first approximative formula to calculate the season's dependent correction factor in the synodic–sidereal transformation of the solar rotation velocity was presented in Graf (1974). Taking into account the very small eccentricity of the Earth's orbit around the Sun the following expression is valid:

$$\boxed{D = 1.0 - e\cos[T + e\sin(T)]} \tag{5.1}$$

where D is the calculated Earth–Sun distance in astronomical units (AU), $e = 0.01675$ is the eccentricity of the Earth's orbit, and $T = (2\pi/365.25) \times (\text{DAY} - \text{PH})$, where DAY is the number of days after January 0 and 0h Ephemeris Time and PH is the time in days (after the same date) of the mean perihelion. We note that there was a typing error in the above formula in the original paper Graf (1974) which is corrected here.

The Earth's daily motion ψ as a function of D is given by

$$\boxed{\psi = \psi_0/D^2} \tag{5.2}$$

where the $\psi_0 = 3547.717$ arc sec is the daily motion of the Earth. Finally we get the solar sidereal angular rotation velocity in deg/day:

$$\boxed{\omega_{\text{sid}} = \omega_{\text{syn}} + \psi} \tag{5.3}$$

where ω_{syn} is the synodic solar angular rotation velocity in deg/day.

5.1.2 Method II: According to Roša et al. (1995) and Brajša et al. (2002)

As we have seen in Chap. 2, the rotation axis of the Sun is defined with the Carrington elements of the solar rotation:

$$i = 7.25^\circ \tag{5.4}$$

$$\Omega = 73.67^\circ + (Y - 1850) \times 0.01396^\circ \tag{5.5}$$

where Y stands for the calendar year for which the value of Ω refers to.

The adopted system of heliographic coordinates is rather arbitrary and the prime meridian is defined with the point on the solar equator which rotates with the sidereal period of 25.38 days. This reference point was chosen to coincide with ascending node on the date of 1 January 1854 at 12:00 UT. In terms of Julian Days this corresponds to the J.D. = 2 398 220.0. The angular distance W of the reference point from the ascending node is defined in the direction of the solar rotation. This is called the direct sense. The angular distance can be calculated from Roša et al. (1995):

$$W = \frac{360^\circ}{25.38}(\text{J.D.} - 2398220.0) \ . \tag{5.6}$$

For any other value of the sidereal period U_{sid} this expression can be written in the following form:

$$W = \frac{360^\circ}{U_{\text{sid}}}(\text{J.D.} - 2398220.0) \ . \tag{5.7}$$

We now recall that the heliographic coordinate system is specified with the heliographic coordinates of the center of the apparent solar disk (L_0, B_0) and with the position angle of the rotation axis P. The following transformation is used according to Waldmeier (1955):

$$\sin B_0 = \sin(\lambda_0 - \Omega)\sin i \ , \tag{5.8}$$

$$\tan(L_0 + W) = \cos i \, \tan(\lambda_0 - \Omega) \ , \tag{5.9}$$

$$P = -\arctan[\cos\lambda_0 \tan\epsilon] - \arctan[\tan i \cos(\Omega - \lambda_0] \ , \tag{5.10}$$

where λ_0 is the apparent longitude of the Sun referred to as the true equinox of the date and ϵ is the obliquity of the ecliptic.

The synodic rotation period of the Sun corresponds to a decrease in L_0 from 360^0. We denote the daily change of the heliographic longitude of the center of the solar disk L_0 with ΔL_0 and the synodic period of the solar rotation U_{syn} is then:

$$\boxed{U_{\text{syn}} = \frac{360^\circ}{\Delta L_0}} \ . \tag{5.11}$$

We suppose now that an observation is performed on some arbitrary date. Let N be the time of the beginning of the observational date and $N + 1$ is the time of the end of that date. This second time coincides with the beginning of the next day. From Eq. (5.9) we get

$$L_0 + W = \arctan[\cos i \tan(\lambda_0 - \Omega)] \tag{5.12}$$

for the time N and

$$\tan(L_0 - \Delta L_0 + W') = \cos i \tan(\lambda_0' - \Omega') \tag{5.13}$$

for the time $N + 1$. We note that the values W', λ_0', and Ω' are related to the time $N + 1$ and ΔL_0 is the daily shift of the parameter L_0, as earlier described.

By subtracting Eqs. (5.12) and (5.13) we get

$$-\Delta L_0 + W' - W = \arctan[\cos i \tan(\lambda_0' - \Omega')] - \arctan[\cos i \tan(\lambda_0 - \Omega)] \ . \tag{5.14}$$

Now from Eq. (5.7) we obtain

$$W' - W = \frac{360°}{U_{\mathrm{sid}}} \ . \tag{5.15}$$

Combining Eqs. (5.14) and (5.15) with (5.7) we obtain the relation between the sidereal (U_{sid}) and synodic (U_{syn}) rotational period of the Sun:

$$\frac{360°}{U_{\mathrm{sid}}} = \frac{360°}{U_{\mathrm{syn}}} + \arctan[\cos i \tan(\lambda_0' - \Omega')] - \arctan[\cos i \tan(\lambda_0 - \Omega)] \ . \tag{5.16}$$

The terms in this equation are expressed in deg/day, while the synodic (U_{syn}) and the sidereal (U_{sid}) periods are expressed in days. The synodic solar rotation period can be measured with observations and then the corresponding sidereal period can be calculated from Eq. (5.16). The values of the longitude of the ascending node, Ω and Ω', can be calculated from Eq. (5.5).

We now describe the calculation procedure for the apparent longitude of the Sun. The geometric mean longitude of the Sun L, referred to as the mean equinox of the date, the Sun's mean anomaly M, and the eccentricity e of the Earth's orbit around the Sun are given by

$$L = 279.69668° + 36000.76892°T + 0.0003025°T^2 \tag{5.17}$$

$$M = 358.47583° + 35999.04975°T - 0.000150°T^2 - 0.0000033°T^3 \tag{5.18}$$

$$e = 0.01675104 - 0.0000418\,T - 0.000000126\,T^2 \ , \tag{5.19}$$

where T is the time measured in Julian centuries of 36,525 ephemeris days from the epoch 1900 January 0.5 ET (Ephemeris Time):

$$T = \frac{\text{J.D.} - 2415020.0}{36525} . \quad (5.20)$$

If the given moment is expressed in the Universal Time (UT), we should add to J.D. the value $\Delta T = \text{ET} - \text{UT}$ given in days. We note that from 1984 on, the relation $\Delta T = \text{TDT} - \text{UT}$ has been used, where TDT is the terrestrial dynamical time. The values of the difference ΔT are tabulated in, e.g., *Astronomical Almanac* and for the several past years the value of ΔT is about one minute.

Finally, the Kepler equation can be solved using the values of M and e, with all angles expressed in degrees:

$$\boxed{E = M + \frac{180^\circ}{\pi} e \sin E ,} \quad (5.21)$$

using the iteration formula:

$$E_{n+1} = E_n + \frac{M - E_n + \frac{180^\circ}{\pi} e \sin E_n}{1 - e \cos E_n} \quad (5.22)$$

to find the eccentricity anomaly E. As a first approximation we use $E = M$. Then the true anomaly f can be obtained using the following expression:

$$\tan\left(\frac{f}{2}\right) = \sqrt{\frac{1+e}{1-e}} \tan\left(\frac{E}{2}\right) . \quad (5.23)$$

The true geometric longitude of the Sun λ referred to the mean equinox of the date is given by

$$\lambda = L + f - M . \quad (5.24)$$

However, to find the apparent longitude of the Sun for the true equinox of the date, λ_0, a further correction is necessary, to take into account the effects of nutation and aberration. A sufficient accuracy is achieved using the formula:

$$\lambda_0 = \lambda - 0.00569^\circ - 0.00479^\circ \sin(259.18^\circ - 1934.142^\circ\, T) . \quad (5.25)$$

Using Eqs. (5.17)–(5.25) the values of the apparent longitude of the Sun for the beginning and the end of the observation day, λ_0 and λ_0', respectively, can be calculated. On the other hand, Eq. (5.5) gives the values of the longitude of the ascending node for the beginning and the end of the observation day, Ω and Ω', respectively. Finally, when these quantities are known, the sidereal rotation period

U_{sid} can be determined from the observed value of the synodic rotation period U_{syn} using Eq. (5.16):

$$\boxed{U_{sid} = U_{syn}\left\{1 + \frac{U_{syn}}{360^\circ}\{\arctan[\cos i\,\tan(\lambda_0' - \Omega')] - \arctan[\cos i\,\tan(\lambda_0 - \Omega)]\}\right\}^{-1}.} \tag{5.26}$$

To make further analysis easier we rewrite the previous Eq. in the following way:

$$U_{sid} = U_{syn}\left\{1 + \frac{U_{syn}}{360^\circ}Q\right\}^{-1}, \qquad Q = Q_1 + Q_2\,, \tag{5.27}$$

with the following substitution:

$$Q_1 = \arctan[\cos i\,\tan(\lambda_0' - \Omega')] - \arctan[\cos i\,\tan(\lambda_0 - \Omega)]\,. \tag{5.28}$$

The quantity Q_2 will be defined and discussed later in this subsection.

When Eqs. (5.27) and (5.28) are used in practical calculations numerical problems might appear. To overcome this problem it should be taken into account to which quadrant the arctan function angle belongs Brajša et al. (2002). These constraints are easier to define if the value of Q_1 is rearranged using the trigonometric identity:

$$\arctan x - \arctan y = \arctan\frac{x - y}{1 + xy}\,. \tag{5.29}$$

Further, we use the following substitutions:

$$x = \cos i\,\tan(\lambda_0' - \Omega')\,, \qquad y = \cos i\,\tan(\lambda_0 - \Omega)\,. \tag{5.30}$$

So, the quantity Q_1 now becomes

$$Q_1 = \arctan\frac{x - y}{1 + xy}\,, \tag{5.31}$$

with the following conditions for Q_2:

$$Q_2 = \begin{cases} 0 \text{ for } xy > -1, \\ \pi \text{ for } x > 0 \text{ and } xy < -1, \\ -\pi \text{ for } x < 0 \text{ and } xy < -1\,. \end{cases} \tag{5.32}$$

We note that the quantity Q should be expressed in degrees before inserting into Eq. (5.27).

Finally, we emphasize that Eq. (5.27), which is valid for rotation periods, can be equivalently expressed by using the angular rotation velocities:

$$\boxed{\omega_{sid} = \omega_{syn} + Q\,,} \tag{5.33}$$

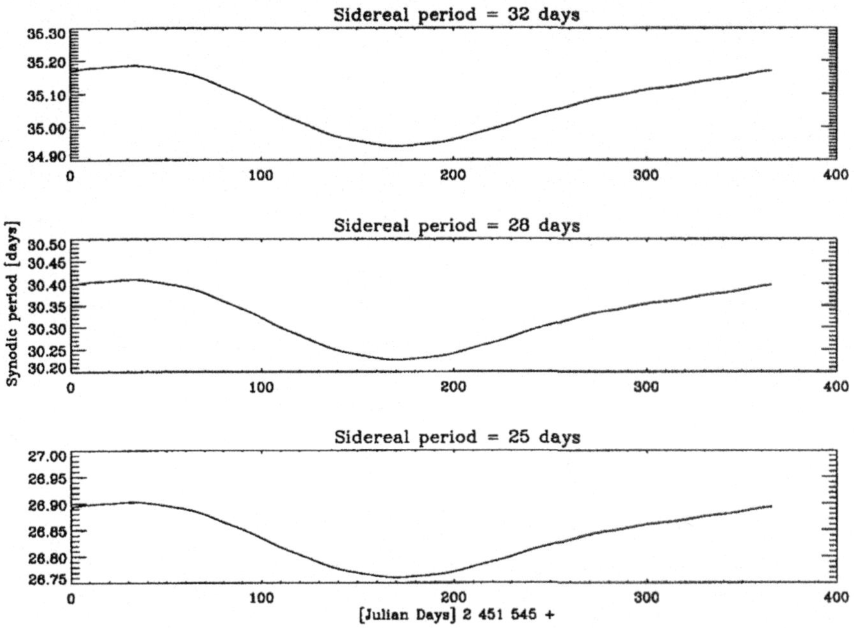

Fig. 5.1 The yearly variation of the synodic rotation period of the Sun calculated using the procedure described in text for three different sidereal periods of the solar rotation. Adapted from Brajša et al. (2002)

where ω_{sid} and ω_{syn} represent the sidereal and the synodic angular solar rotation velocity, respectively. For completeness, note that the angular rotation velocity ω and the rotation period U are generally related by

$$\boxed{\omega = \frac{360^\circ}{U}\,.} \tag{5.34}$$

The quantity Q is the daily angular difference between the two rotation velocities and is expressed in the same units as ω_{sid} and ω_{syn}. As we have already seen in this chapter, this difference is season's dependent and its value changes during the year. An example of this variation is presented in Fig. 5.1 for three different sidereal periods of the Sun.

5.1.3 *Method III: According to Wittmann (1996)*

An easier and more direct procedure for synodic–sidereal transformation, than described in the previous subsection, is described in Wittmann (1996). However, it is not fully correct, as we will see later in this section. In that analysis we begin with the angular velocity difference $\Delta\omega$:

$$\Delta\omega = \omega_{\rm sid} - \omega_{\rm syn} = \omega_{\rm orb} \cos\Phi \;, \tag{5.35}$$

which should be added to the observed synodic rotation velocity, $\omega_{\rm syn}$, to get the true sidereal rotation velocity $\omega_{\rm sid}$. The angular velocity of the Earth while moving on its orbit around the Sun, $\omega_{\rm orb} \approx 1$ deg day^{-1}, is not a constant due to the eccentricity of the Earth's orbit which amounts to $e \approx 0.0167$. The variation of $\omega_{\rm orb}$ during the year is about $\pm 1.6\,\%$. The quantity Φ denotes the angle between the ecliptic plane and the solar rotation vector, which is assumed to be orthogonal to the solar rotation axis.

Now we take into account Kepler's second law of planetary motions, which states that the planet sweeps the same area in the same time interval:

$$\boxed{\Delta f = r^2\, \omega_{\rm orb} = \text{constant} \;,} \tag{5.36}$$

where r is the (variable) distance between the Earth, where the observer is located, and the Sun. For the Earth in modern times this constant is $\Delta f = 0.9856$ deg day^{-1} AU^2, where AU denotes the astronomical unit.

Finally, we can rewrite Eq. (5.35):

$$\boxed{\Delta\omega = \omega_{\rm sid} - \omega_{\rm syn} = \frac{0.9856 \cos\Phi}{r^2} \;.} \tag{5.37}$$

The distance r can be calculated according to Meeus (1991), while the quantity Φ can be calculated according to Very (1897):

$$\boxed{\Phi = \arctan[-\tan i \cos(\lambda_0 - \Omega)] \;,} \tag{5.38}$$

where i is the angle between the ecliptic plane and the equatorial plane of the Sun, λ_0 is the apparent ecliptic longitude of the Sun, and Ω is the ecliptic longitude of the ascending node of the solar equator. Note that in the paper Wittmann (1996) appeared a couple of typing errors, which are corrected here (see also a comment in the next subsection).

5.1.4 *Method IV: According to Skokić et al. (2014); A Comparison of Various Methods*

In the last several decades various methods for the calculation of the solar sidereal rotation period from the synodic one were presented, as described in the previous subsections of this chapter. Hereafter, we discuss the validity of those methods by comparing them with the best available ephemeris data. We will show that the relations by Graf (1974) and Roša et al. (1995) are in general agreement with the ephemeris calculations, while that of Wittmann (1996) shows a discrepancy of almost 0.01° day^{-1}.

Sidereal solar rotation velocity, $\omega_{\rm sid}$, is usually calculated by adding the mean orbital angular velocity of the Earth, $\overline{\omega}_{\rm Earth}$, to the measured synodic velocity $\omega_{\rm syn}$:

$$\boxed{\omega_{\rm sid} = \omega_{\rm syn} + \overline{\omega}_{\rm Earth}\ .} \tag{5.39}$$

For the $\overline{\omega}_{\rm Earth}$ a value of 0.9856° day^{-1} is usually used. Graf (1974) pointed out that this approach leads to periodic errors of up to 0.15° day^{-1}, because it does not include the effects of ellipticity of the Earth's orbit and inclination of solar rotation axis to the ecliptic, and suggested a new relation which takes into account these effects. His method was later improved by Roša et al. (1995), with further corrections in Brajša et al. (2002):

$$\omega_{\rm sid} = \omega_{\rm syn} + \left\{\arctan[\cos i \tan(\lambda_0' - \Omega')] - \arctan[\cos i \tan(\lambda_0 - \Omega)]\right\}\ , \tag{5.40}$$

where λ_0 is the apparent longitude of the Sun referred to as the true equinox of the date, Ω is the longitude of the ascending node of the solar equator on the ecliptic, and i is the inclination of the solar equator to the ecliptic. Primed variables denote the values one day after the non-primed values. As we have seen earlier in this chapter, Roša et al. (1995) also gave the approximate equations to calculate λ_0 and Ω, without the need for using published ephemerides.

Equation (5.40) has two disadvantages. First, the attribution of the correct quadrant should be taken into account when calculating the arctangent functions. Otherwise the discontinuities could emerge and lead to wrong results (see Fig. 5.3). This problem was discussed in Brajša et al. (2002). The second difficulty is that the calculation of λ_0 and Ω for two consecutive days is needed. A simpler method which avoids this was proposed by Wittmann (1996):

$$\omega_{\rm sid} = \omega_{\rm syn} + \omega_{\rm Earth} \cos\Phi\ , \tag{5.41}$$

where $\omega_{\rm Earth}$ is the instantaneous orbital angular velocity of the Earth and Φ is the angle between the pole of the ecliptic and the solar rotation axis orthographically projected back onto the solar disk (see Fig. 5.2). This angle can be calculated from Very (1897):

$$\tan\Phi = \tan i \cos(\lambda_0 - \Omega)\ . \tag{5.42}$$

We note that $\omega_{\rm Earth}$ can be found with sufficient accuracy using Kepler's second law (e.g., Goldstein (1983), Eq. (3-53)):

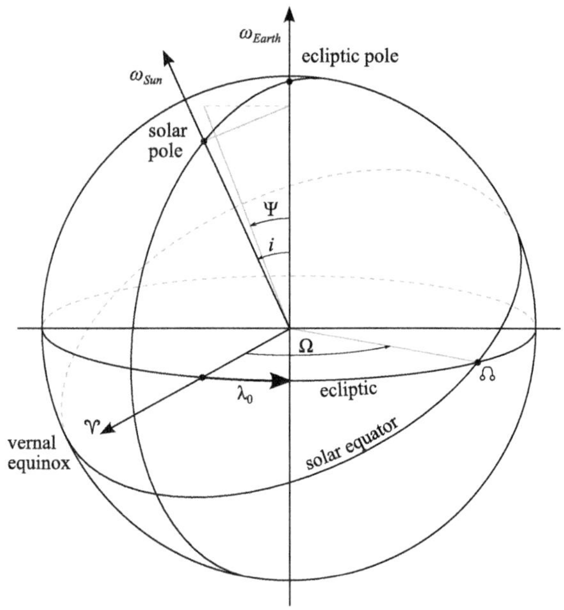

Fig. 5.2 Definition of the angles used in the analysis. The Earth is positioned towards the reader. Note that the angle Φ is here denoted with Ψ. Adapted from Skokić et al. (2014)

$$\boxed{\omega_{\text{Earth}} = \frac{\overline{\omega}_{\text{Earth}}}{r^2}\ ,} \tag{5.43}$$

where r is the distance between the Earth and the Sun, expressed in AU. We also point out that a typing error in Wittmann (1996) occurred, where on several occasions a value of $0.9865°\ \text{day}^{-1}$ instead of the correct value $0.9856°\ \text{day}^{-1}$ was printed for $\overline{\omega}_{\text{Earth}}$.

To investigate the validity of the given relations, we used ephemeris data from JPL Horizons system[1] (for further details and references see Skokić et al. (2014)), to calculate the heliographic longitude of the Earth, L_0, and instantaneous sidereal-synodic correction by using the following relation:

$$\boxed{\omega_{\text{sid}} - \omega_{\text{syn}} = \omega_C + \frac{dL_0}{dt}\ ,} \tag{5.44}$$

where $\omega_C = 14.1844°\ \text{day}^{-1}$ is the Carrington rotation rate. This correction was applied to the fixed sidereal rotation velocity, ω_{sid}, for which we also used the Carrington rate with a period of 25.38 days, to get the reference synodic velocity. Also, the synodic velocities were calculated by Eqs. (5.40) and (5.41), and the results were compared in Fig. 5.3. It can be seen that method by Roša et al. (1995) agrees well with the JPL ephemeris data, while that of Wittmann (1996) shows a small discrepancy.

[1] http://ssd.jpl.nasa.gov/?horizons.

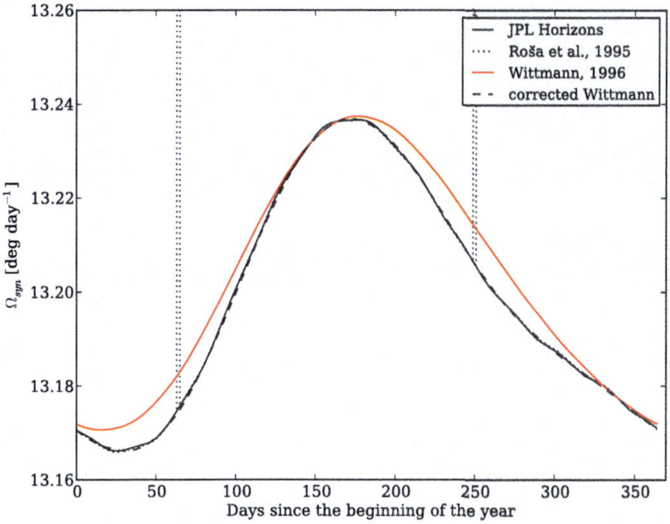

Fig. 5.3 Synodic velocity as a function of time as calculated by different methods for a fixed sidereal velocity of 14.1844° day^{-1}. The method of Wittmann (1996) gives slightly different results from the others which show almost the same behavior. The discontinuities in the method of Roša et al. (1995), which take place when arctangent functions are not handled properly, are also shown. Adapted from Skokić et al. (2014)

A careful check of Eq. (5.41) shows that a factor is missing that re-projects orthographically the projected orbital angular velocity of the Earth back to the solar rotation axis. If this is done, one gets the following relation:

$$\boxed{\omega_{\text{sid}} = \omega_{\text{syn}} + \omega_{\text{Earth}} \frac{\cos^2 \Phi}{\cos i}\ .} \tag{5.45}$$

As can be seen from Fig. 5.3, this corrected Wittmann relation agrees well with the JPL ephemeris data and does not result in discontinuities. In Fig. 5.4 we compare residual errors for different methods. Method by Wittmann shows a maximum error of $\sim 0.008°$ day^{-1}, while others are about ten times smaller. The residual error in the method by Roša et al. (1995) is due to the usage of approximate formulae for the calculation of Earth's orbital elements, while one-month variations in the corrected Wittmann method are most probably due to the effect of the Moon which was neglected in Eq. (5.43) for the calculation of the Earth's orbital angular velocity. When full JPL ephemeris are used and ω_{Earth} is calculated from Earth's linear orbital velocity or from the daily change of λ_0, these variations disappear.

Equation (5.45) can also be derived analytically. Since instantaneous synodic–sidereal correction is simply given by the temporal derivative of the Earth heliographic longitude, Eq. (5.44), there are two ways to calculate it. The first procedure is by estimating L_0 at two consecutive times, which leads to Eq. (5.40). The second

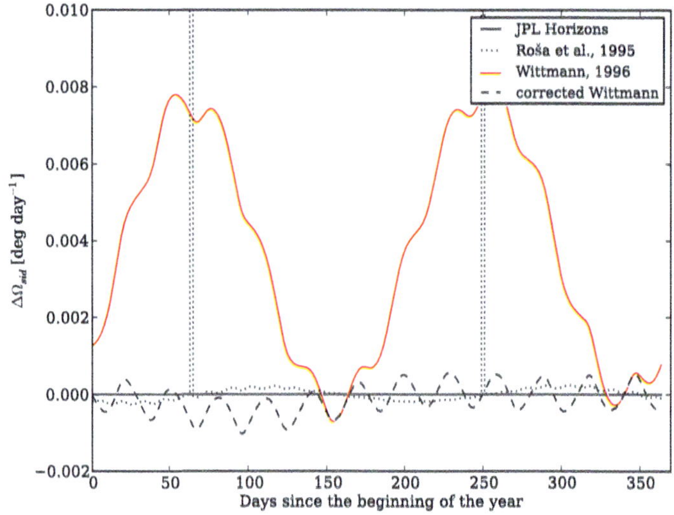

Fig. 5.4 Residual errors as a function of time for different calculation methods of sidereal velocity. Adapted from Skokić et al. (2014)

procedure is by differentiating L_0. We start with the relation for L_0 (e.g., Meeus (1991), Roša et al. (1995)):

$$\tan(L_0 + W) = \tan(\lambda_0 - \Omega)\cos i \ , \tag{5.46}$$

where W is the angular distance of the Carrington meridian from the ascending node. The temporal derivative of L_0 is

$$\frac{dL_0}{dt} + \frac{dW}{dt} = \frac{d}{dt}\{\arctan[\tan(\lambda_0 - \Omega)\cos i]\} \ . \tag{5.47}$$

By using Eq. (5.44) and $\omega_C = dW/dt$, instead of the left-hand side term, and calculating the derivative of the right-hand side term (neglecting $d\Omega/dt \approx 50$ arcsec/year), we get

$$\omega_{\text{sid}} - \omega_{\text{syn}} = \frac{\cos i}{[1 + \tan^2(\lambda_0 - \Omega)\cos^2 i]\cos^2(\lambda_0 - \Omega)} \frac{d\lambda_0}{dt} \ . \tag{5.48}$$

The last term on the right-hand side, $d\lambda_0/dt$, represents Earth's angular orbital velocity, ω_{Earth}. Sorting out the denominator we get

$$\boxed{\omega_{\text{sid}} - \omega_{\text{syn}} = \omega_{\text{Earth}} \frac{\cos i}{\cos^2(\lambda_0 - \Omega) + \sin^2(\lambda_0 - \Omega)\cos^2 i} \ ,} \tag{5.49}$$

while converting sine to cosine and rearranging the denominator lead back to Eq. (5.45).

We can conclude that several methods for the calculation of the sidereal solar rotation velocity from the observed synodic velocity were introduced. Using state-of-the-art ephemeris data, we found that the method by Roša et al. (1995), Eq. (5.40), agrees well with JPL ephemeris calculations, while the method by Wittmann (1996), Eq. (5.41), shows some discrepancy and was later corrected with Eq. (5.45). Although the error in using incorrect Eq. (5.41) is only of the order 0.01° day^{-1}, this difference can be important in studies of all phenomena related to solar rotation and their time variation. This is especially important when the measured values have the same order of magnitude.

The precision of the synodic–sidereal correction depends on the precision of L_0 and, consequently, λ_0, and numerical evaluation of their derivatives. Ephemeris systems and almanacs usually list L_0 with precision of 0.01°. Since synodic velocity is a temporal derivative of L_0, the precision of synodic–sidereal correction should only be affected by time variable terms which are usually given with much higher precision. The ultimate precision on the synodic–sidereal correction is also related to the uncertainties of the Carrington elements (i and Ω) (e.g., Balthasar et al. (1986, 1987), Beck and Giles (2005), see also one of the previous chapters). Thus, by using Eqs. (5.40) and (5.45), with the best available ephemeris, maximum precision is reached. If approximate formulae are used as given in Roša et al. (1995) and Eq. (5.43), then the precision of synodic–sidereal correction is of the order of 0.001° day^{-1}.

We advise to use any of Eq. (5.40), Roša et al. (1995), or (5.45) (Wittmann corrected). An even better method would be transforming the observed central meridian distances to a coordinate system which is defined and fixed with solar rotation axis and then directly calculating the sidereal velocities without the need of synodic to sidereal transformation. Finally, we note that the original Wittmann equation (Eq. (5.41)) would give substantially larger errors for greater inclinations than in the Sun–Earth case.

5.2 Height Correction in the Solar Rotation Determination

If the observed tracers are located or extended above the solar photosphere, then in fact their projected positions on the photosphere are measured, when observing in the white light. This is illustrated in Fig. 5.5 where it is clearly seen how the measured, projected position differs from the footpoint position of the observed tracer. It is important to note that both, the latitude b and the longitude, expressed through the central meridian distance (CMD), are changed. The amount of the effect depends on the location of the observed tracer on the solar disk; the effect is larger towards the solar limb, both in the east–west and north–south directions. These projection effects have a significant influence on the solar rotation determination, leading to the false velocity determination. We can visualize this effect in a simple way described in

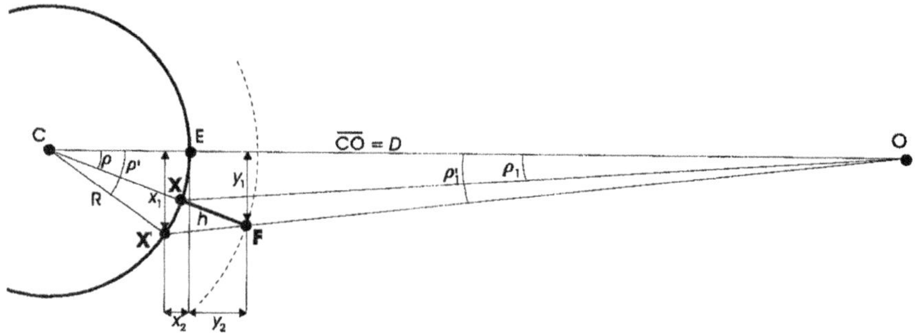

Fig. 5.5 Geometrical concept of the projection effect due to the height of the tracer above the photosphere. O denotes the observer, C the center of the Sun, F the observed tracer, X its footpoint, X' the projection of the tracer on the solar disk, and E the sub-Earth point. Adapted from Roša et al. (1998)

the following. Consider a filament, extended in the solar chromosphere and corona, whose apparent position is measured during its passage across the solar disk at medium latitudes. The measured CMD values have larger negative values on one side from the central meridian and larger positive values on the other side. When the solar rotation velocity is determined using these apparent values, we obviously get too large apparent velocities, since a larger distance is divided by the same elapsed time difference. These rotation velocity results are also associated with displaced, false latitudes.

Finally we note that projection effects could lead to false rotation results also for tracers seen below the photosphere. This is the case of the Wilson effect for sunspots, which apparently changed the position of a sunspot close to the solar limb.

5.2.1 Description of the Method

A method for the simultaneous determination of the corrected solar rotation velocity and height of the tracers was derived in Roša et al. (1998). The method is based on spherical trigonometry, the least-square fit was applied, and the error analysis was performed. A detailed description of the method is given in Roša et al. (1998) and here we reproduce only the final equations for the application of the method. We note that we use here the notation from Brajša et al. (2004), which is slightly different from Roša et al. (1998), and all quantities are defined at appropriate places in the text. Some symbols are specific only to this section and have different meanings in other sections.

Quantities denoted with the asterisk (*) are the measured ones and without asterisk the corrected ones. First we introduce the parameter β, which connects the height parameter $\epsilon = h/R$, where h is the height above the solar surface and R the solar radius, with the observed latitude, ψ^*, and corrected (true) latitude of the tracer, ψ

$$\beta = (1+\epsilon)\frac{\cos\psi}{\cos\psi^*} . \tag{5.50}$$

This is Eq. (17b) in Roša et al. (1998), which can be approximatively written as

$$\beta = \frac{\sqrt{(1+\epsilon)^2 - \sin^2\psi^*}}{\cos\psi^*} = \text{constant} , \tag{5.51}$$

for the case $B_0 \approx 0$, where B_0 is the heliographic latitude of the solar disk center. This is Eq. (20b) in Roša et al. (1998). The relative deviation of the parameter β from a constant value due to the change of the projected heliographic latitude is less than 2% even in the most inconvenient cases when B_0 reaches the maximum value and the tracer is close to the limb (Roša et al. 1998).

Now we connect the parameter β with the mean observed central meridian distance λ^*, and corresponding observed rotation velocity ω^*, and corrected rotation velocity ω, of the measured tracer

$$\omega_i^* = \frac{\sqrt{\beta^2 - \sin^2\lambda_i^*}}{\cos\lambda_i^*}\omega . \tag{5.52}$$

This is Eq. (21d) in Roša et al. (1998).

Further, the corrected rotation velocity ω and the parameter β can be calculated by

$$\boxed{\omega^2 = \frac{ab - cd}{e - d^2}} , \tag{5.53}$$

$$\boxed{\beta^2 = \frac{b - d + c/\omega^2}{b}} , \tag{5.54}$$

which are Eqs. (33a) and (33b) in Roša et al. (1998), respectively. The following abbreviations were introduced:

$$a = \sum_{i=1}^{N} \omega_i^{*2} , \tag{5.55}$$

$$b = \sum_{i=1}^{N} \frac{1}{\cos^4\lambda_i^*} , \tag{5.56}$$

$$c = \sum_{i=1}^{N} \frac{\omega_i^{*2}}{\cos^2\lambda_i^*} , \tag{5.57}$$

$$d = \sum_{i=1}^{N} \frac{1}{\cos^2 \lambda_i^*} \,, \tag{5.58}$$

$$e = \sum_{i=1}^{N} \frac{N}{\cos^4 \lambda_i^*} \,, \tag{5.59}$$

where the summation refers to 10 deg latitude bins and N is the number of measured pairs of velocities ω_i^* and mean central meridian distances λ_i^*, in each bin.

Finally, the true height of the tracer is calculated by

$$\boxed{h = R \left(\sqrt{\beta^2 \cos^2 \psi^* + \sin^2 \psi^*} - 1 \right) \,,} \tag{5.60}$$

where ψ^* is the mean latitude value for each latitude bin

$$\psi^* = \frac{1}{N} \sum_{i=i}^{N} \psi_i^* \,. \tag{5.61}$$

These two expressions are Eqs. (34) and (35) in Roša et al. (1998). Equation (5.60), as well as Eq. (5.51), hold in the approximation $B_0 \approx 0$. The mean value of the measured rotation velocities is calculated by

$$\omega^* = \frac{1}{N} \sum_{i=i}^{N} \omega_i^* \,, \tag{5.62}$$

for each latitude bin. The corrected latitude ψ is calculated by

$$\boxed{\cos \psi = \frac{\beta \cos \psi^*}{\sqrt{\beta^2 \cos^2 \psi^* + \sin^2 \psi^*}} \,.} \tag{5.63}$$

This formula is obtained by combining Eqs. (5.50) and (5.51).

5.2.2 *Application of the Method*

In the previous subsection, we described the method for simultaneous determination of the corrected rotation velocity and height of the tracers. The method also provides the determination of the corrected (true) latitude of the observed object, which is important when the dependence of the rotation velocity on the latitude is analyzed. In this subsection, we briefly summarize the main results of three successful applications of this method, for three different tracers: Hα filaments, low brightness temperature regions, and coronal bright points.

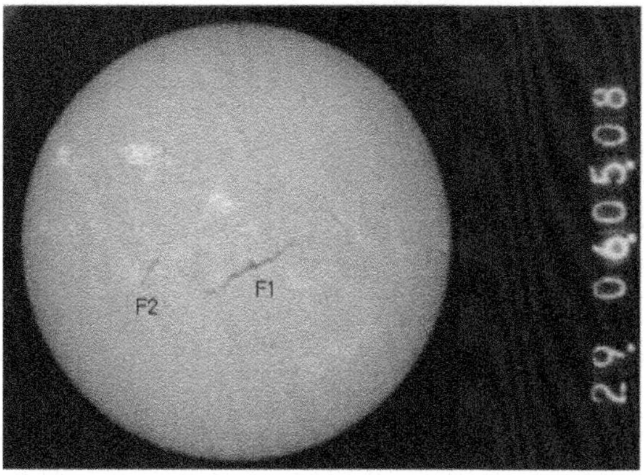

Fig. 5.6 The full-disk solar filtergram taken on 29 May 1993 at 06:05 UT in Hα at Kanzelhöhe Observatory for solar and environmental research (Institute of Physics, University of Graz, Austria). The analyzed filaments are denoted with F1 and F2. Adapted from Vršnak et al. (1999)

In the first case prominences, seen as filaments on the disk, were analyzed on Hα full-disk solar images taken at the Kanzelhöhe Observatory for solar and environmental research, Institute of Physics, University of Graz, Austria (Vršnak et al. 1999). Several images per day were taken, enabling precise determination of the solar rotation velocity. An example of such full-disk solar image is given in Fig. 5.6, where the two filaments, whose positions were measured, are denoted as F1 and F2. In Fig. 5.7 results for three segments of the filament F1 are presented, as the measured rotation speed as a function of the measured central meridian distance (CMD). The dots in the panels would be aligned parallel to the x-axis if the projection effects would not be present. However, since the observed tracers extend high in the solar atmosphere and the projection effects are significant, the apparent rotation rate speeds up as the CMD approaches the solar limb, which is clearly seen in this figure. For each filament segment the corrected rotation velocity ω and the parameter β were calculated, according to Eqs. (5.53) and (5.54), respectively, and the measured rotation velocities were fitted as a function of the measured CMD using Eq. (5.52). The fitted curves are presented in each panel in Fig. 5.7. The output results are also given inserted in each panel of Fig. 5.7. It is important to note that in further analysis the corrected values of the rotation rate and the latitude are used. Finally, the heights of the measured filament segments are in the range from about 53,000 to 92,000 km.

In the second example, the authors Brajša et al. (2009) analyzed daily full-disk radio maps of the Sun recorded at the wavelength of 8 mm (corresponding to the frequency of 37 GHz). The measurements were performed during several years at the Metsähovi Radio Observatory, Helsinki University of Technology, Finland, and the so-called low brightness temperature regions (LTRs) were identified and traced in sequences of successive maps. These objects have lower brightness temperatures, i.e., intensities, than the quiet Sun level at the measurement frequency and are often identified as signatures of Hα filaments observed in the mm wavelength range. Data reduction was performed for 10 deg latitude bins and for each latitude bin the corresponding height was obtained. The analysis of more than 1000 LTRs distributed in 10 deg latitude bins in the latitude range 0–50 deg resulted in the average height

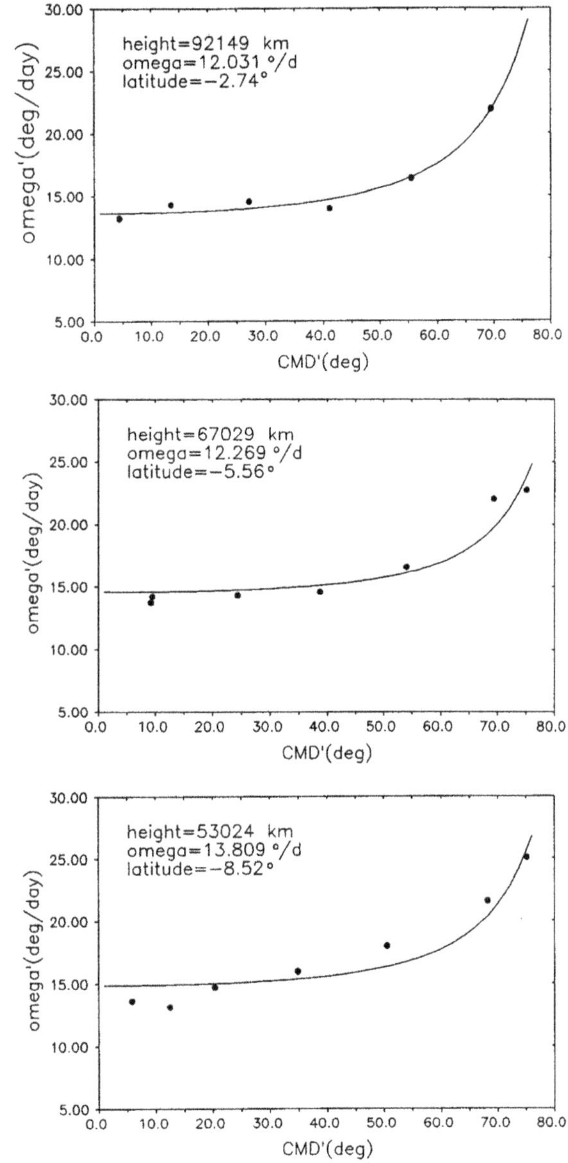

Fig. 5.7 Measured rotation velocity (omega') as a function of the central meridian distance (CMD') for three different segments of the filament F1 (see Fig. 5.6). The latitude of each segment is approximately the same. Adapted from Vršnak et al. (1999)

$h = 45{,}600$ km, in general agreement with the prominence/filament height above the photosphere.

Finally, in the third example, we briefly describe the analysis of the height correction in measurements of the solar differential rotation determined by coronal bright points Brajša et al. (2004). Similar as in the previous case, measured rotation velocities are divided into 10 deg latitude bins and the results are presented in Fig. 5.8. In

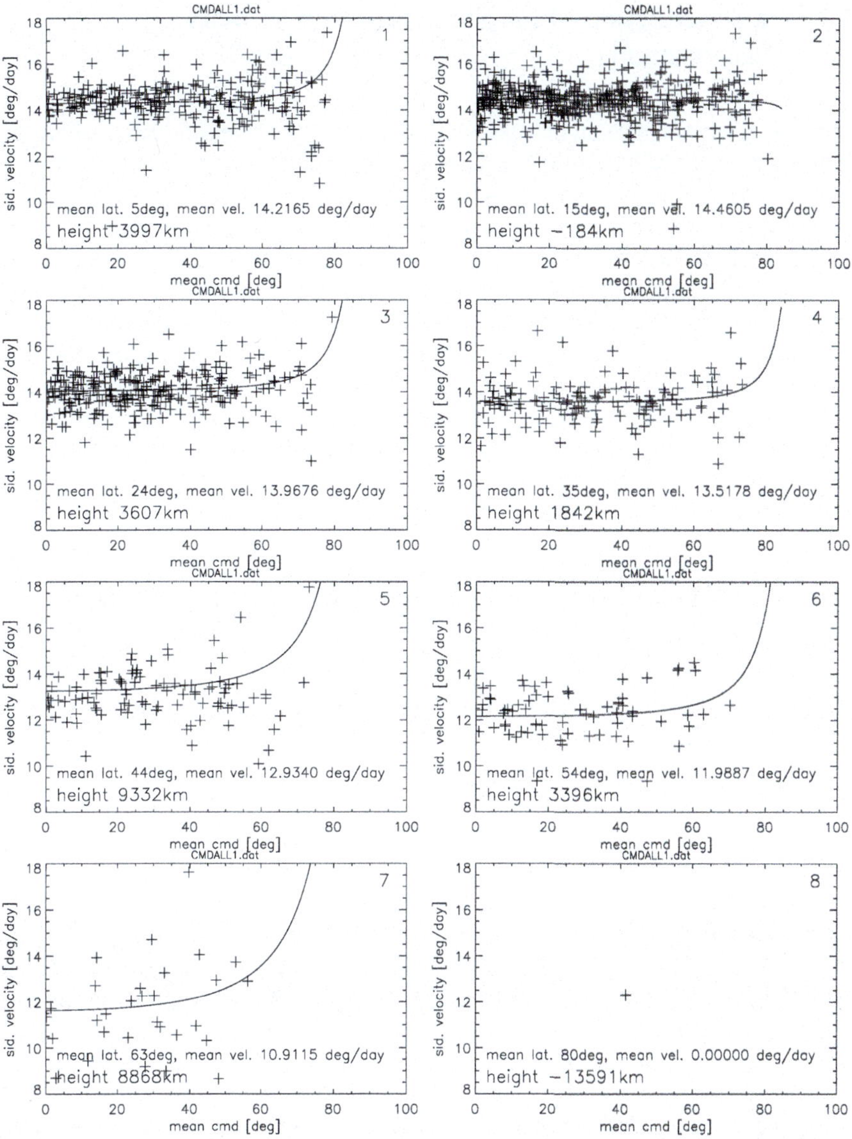

Fig. 5.8 Similar to Fig. 5.7 for about 1000 coronal bright points measurements distributed in 10 deg latitude bins. Full-disk solar images taken in the 1998/1999 period with the EIT instrument (28.4 nm filter) onboard the SOHO spacecraft were used. Adapted from Brajša et al. (2004)

each panel, the calculated height of the tracers for the corresponding latitude bin is given. The calculated height of coronal bright points is on the average 8,000–12,000 km above the solar photosphere.

5.3 Data Reduction of the Solar Rotation Velocity Measurements

After we described in detail how the solar rotation velocities as a function of the latitude are determined from measurements, in this section we continue with a description of the procedure on how to deal further with the obtained observational data. In the following subsections, we treat the four subtopics related to the data reduction of solar rotation velocity measurements. First, we present how the solar differential rotation is mathematically represented with solar rotation parameters. Then we give a short description of the least-square method, which is used in various stages of the data reduction. Further, we point out the importance of the so-called cross-talk between the solar rotation parameters, which can lead to a not unique representation of the solar differential rotation. Finally, the problem of filtering out the erroneous data outliers is briefly addressed. The possibilities of how the mentioned issues could be overcome are also discussed.

5.3.1 *Mathematical Representation of the Solar Differential Rotation*

As we have already mentioned in Chap. 1 the solar differential rotation is most often represented by the following formula:

$$\boxed{\omega(b) = A + B\sin^2 b + C\sin^4 b\ ,} \tag{5.64}$$

where ω is the solar angular rotation velocity, expressed in deg/day, b the heliographic latitude, expressed in deg, and A, B, and C the solar rotation parameters. The solar rotation parameter A represents the rotation velocity at the solar equator, while the parameters B and C represent the gradient of the solar differential rotation, i.e., the dependence on the latitude. The parameter C is important for high solar latitudes and can be neglected for low latitude tracers, such as sunspots. The solar angular rotation velocity can be expressed in deg/day, in μrad/s, or in nHz. Additionally, it is also important to know the orbital speed of the Sun, expressed in m/s, which is dependent on the distance from the rotation axis, which is further related to the latitudinal position of the part of the solar surface under consideration.

The solar angular rotation velocity ω can easily be transformed into period of rotation in days. If ω is expressed in deg/day, the rotation period in days is simply given by

$$\boxed{P = \frac{360^\circ}{\omega}\ .} \tag{5.65}$$

5.3.2 The Least-Square Method

There is a wide choice of general literature about the least-square method, e.g., Anderson (2011), Press et al. (2002), and Ivezić et al. (2014). The least-square method is often presented in a variety of notations and here we use the notation from Press et al. (2002). The analysis is described following partly Press et al. (2002), in a slightly simplified form.

We start with a set of experimental data, collected from astronomical observations. Let us consider a series of N pairs of measured quantities (x_i, y_i). We would like to describe the data with a straight line and assume a linear relationship:

$$\boxed{y(x) = a + bx\ ,} \tag{5.66}$$

where a and b are parameters to be determined. These parameters, also called coefficients, describe the experimental data in the best way. So, we apply the least-square method to fit the data with a straight line. This analysis is also called linear regression.

Next we define the chi-square merit function:

$$\chi^2(a, b) = \sum_{i=1}^{N}(y_i - a - bx_i)^2\ . \tag{5.67}$$

To determine the parameters a and b this Eq. should be minimized. This is achieved when derivatives of $\chi^2(a, b)$ vanish with respect to a and b:

$$\frac{\partial \chi^2}{\partial a} = -2\sum_{i=1}^{N}(y_i - a - bx_i) = 0 \tag{5.68}$$

$$\frac{\partial \chi^2}{\partial b} = -2\sum_{i=1}^{N}x_i(y_i - a - bx_i) = 0. \tag{5.69}$$

Now we introduce the following useful abbreviations:

$$S_x \equiv \sum_{i=1}^{N} x_i \tag{5.70}$$

$$S_y \equiv \sum_{i=1}^{N} y_i \tag{5.71}$$

$$S_{xx} \equiv \sum_{i=1}^{N} x_i^2 \tag{5.72}$$

$$S_{xy} \equiv \sum_{i=1}^{N} x_i y_i. \tag{5.73}$$

Using these substitutions Eqs. (5.68) and (5.69) can be written as follows:

$$S_y = aN + bS_x \tag{5.74}$$

$$S_{xy} = aS_x + bS_{xx}, \tag{5.75}$$

respectively. This system of two equations with two unknowns can be solved as follows:

$$\Delta \equiv NS_{xx} - (S_x)^2 \tag{5.76}$$

$$\boxed{a = \frac{S_{xx}S_y - S_x S_{xy}}{\Delta}} \tag{5.77}$$

$$\boxed{b = \frac{NS_{xy} - S_x S_y}{\Delta}.} \tag{5.78}$$

The last two Eqs. give the best-fit model parameters a and b.

Taking into account the propagation of errors it can be shown that the standard errors, M_a and M_b, of the parameters a and b have the following form:

$$\boxed{M_a = \frac{\chi}{\sqrt{N-2}}\sqrt{\frac{S_{xx}}{\Delta}}} \tag{5.79}$$

$$\boxed{M_b = \frac{\chi}{\sqrt{N-2}}\sqrt{\frac{N}{\Delta}}.} \tag{5.80}$$

Finally, one measure of the quality, i.e., goodness of the fit is the correlation coefficient, r_{ab}:

$$\boxed{r_{ab} = \frac{-S_x}{\sqrt{NS_{xx}}}.} \tag{5.81}$$

The correlation coefficient is a number between -1 and 1, and its positive value indicates that the errors in the fits coefficients a and b probably have the same sign, while the negative value is an indication that these errors are anticorrelated, probably having opposite signs. Correlation coefficient is an important measure if the functional dependence of the measured quantities, Eq. (5.66), really exists and how good the best-fit parameters, a and b, describe the data from measurements or observations.

In this subsection, we have described the fitting procedure of a straight line, which is defined by the linear relation (5.66), which is the polynomial of the first degree. A similar procedure can be derived also for the polynomials of a higher degree. In the context of the solar rotation determination, the most important polynomials are those of the first degree (the procedure is described here) and of the second degree having the form:

$$y(x) = a + bx + cx^2 \ , \tag{5.82}$$

where a, b, and c are the parameters to be determined.

We note that in many computer program languages, such as the Interactive Data Language (IDL) and Python, widely used in astronomy, the routines for application of the least-square fit method are included in the standard packages. The fit parameters, their standard errors, and correlation coefficients can easily be calculated for the input data and defined polynomial degree.

5.3.3 Application of the Least-Square Method in the Solar Rotation Determination

There is a variety of very useful applications of the least-square method in the solar rotation analysis. Here we describe two important examples.

As the first example, let us consider the simple passage method, a subclass of the tracer method, described in previous chapter. The observed quantities are $y_i = CMD_i$, where CMD is abbreviated for the central meridian distance, and $x_i = t_i$, where t denotes the time; see Fig. 4.15. We now apply the linear least-square fit to the function $y(x) = a + bx$, or $CMD(t) = a + bt$. We must have at least three pairs of measurements: $N \geq 3$. The best-fit parameter $b = \omega$ is then the angular rotation velocity, expressed in, e.g., deg/day. We can also easily calculate the corresponding standard error of this parameter, $M_b = M_\omega$ and the correlation coefficient r. These two error estimates can help to decide if the obtained rotation velocity is real at all. Namely, due to the possible false tracer identification it can happen that erroneous rotation velocities are determined. Moreover, in this way, a quantitative criterion can be established for accepting or rejecting a particular rotation velocity obtained from measurements. Finally, we note that the fit parameter a is not important in this case.

In the second example, we use the least-square method for the determination of the solar differential rotation parameters. First, we point the attention of the reader to the fact that some symbols (letters) have different meanings in different contexts and this is extremely hard to avoid. So, b denotes the solar latitude in Eq. (5.46) and the fitting parameter in Eq. (5.66). Further, let us assume that we have a series of observational data, pairs of the solar angular rotation velocities, ω_i, measured at the latitudes, b_i. Now we come back to the formula for the solar differential rotation, Eq. (5.46), and repeat it here:

$$\omega(b) = A + B\sin^2 b + C\sin^4 b \ , \tag{5.83}$$

where the coefficients A, B, and C are the solar differential rotation parameters. Using the substitution $x = \sin^2 b$ this equation becomes a polynomial of the second degree. If only the low and middle solar latitudes are covered by observations, the parameter C can be neglected and with $C = 0$ Eq. (5.83) becomes a polynomial of the first degree. In both cases, the least-square method can be applied and the parameters A, B, and C can be determined with corresponding standard errors.

5.3.4 *Calculation of the Mean Values and Corresponding Standard Errors*

Let us assume that we have a series of rotation velocity measurements in a latitude bin. The width of latitude bins is typically 2, 5, or 10 deg, according to the number of individual velocity values and the latitudinal coverage of observations. Now we answer the following question. What is the mean rotation velocity value and corresponding error for a specific latitude bin. For completeness, we repeat here the general, well-known formulae. The mean value $\bar{x}$ of a set of measured quantities x_1, x_2, x_3, ... x_N is

$$\boxed{\bar{x} = \frac{\sum_{i=1}^{N} x_i}{N}} \ , \tag{5.84}$$

where N is the number of measurements. The corresponding standard deviation m is given by

$$\boxed{m = \sqrt{\frac{\sum_{i=1}^{N} (\bar{x} - x_i)^2}{N-1}}} \ . \tag{5.85}$$

Finally, we would like to calculate the corresponding standard error M, also called the mean quadratic error of the mean value:

$$\boxed{M = \frac{m}{\sqrt{N}} = \sqrt{\frac{\sum_{i=1}^{N} (\bar{x} - x_i)^2}{N(N-1)}}} \ . \tag{5.86}$$

Please note that this simple procedure is valid for a variety of measured quantities. The mean value of a measured data set can be represented with $\bar{x} \pm M$. The standard errors, accompanied with the mean values, are important when the results of different observing data sets are compared and the statistical significance of their possible difference is investigated.

5.3.5 *Cross-Talk Between the Solar Differential Rotation Parameters and Orthogonal Functions*

One of the main problems in the analysis of solar differential rotation is the question whether the mathematical representation of the differential rotation profile is unique. We repeat the differential rotation formula:

$$\omega(b) = A + B\sin^2 b + C\sin^4 b \ , \tag{5.87}$$

where the parameter A represents the angular rotation rate at the solar equator and the parameters B and C describe the differential character of the solar rotation. Since the angular rotation rate is decreasing with increasing latitude b, one would expect that both parameters B and C have a negative sign. However, in reduction of actual data, it can happen that one of the two parameters, B or C, has the negative sign, as expected, but that the other one has a positive sign. Still, the differential rotation profile can retain the usual behavior, i.e., decreasing the angular rotation rate with increasing latitude. On the other hand, we can get rather similar differential rotation profiles for various combinations of negative parameters B and C. This is the phenomenon of the so-called cross-talk between the solar rotation parameters and we see that the mathematical representation of the solar differential rotation using the above Eq. is not unique, which makes a comparison between various observational results rather difficult.

There are several possibilities to overcome this problem. A simple solution is to put $C = 0$. Then we have only the single parameter B which defines the differential profile of the solar rotation. However, the assumption $C = 0$ can be justified only if the observations cover low and middle latitudes, with no data at high latitudes and in polar regions. This is typical for, e.g., sunspots.

Another possibility to avoid the cross-talk problem is to adopt the additional assumption $B = C$ in the three-parameter fit, as proposed in Scherrer et al. (1980). We use again the substitution $x = \sin^2 b$ and now Eq. (5.87) becomes

$$\boxed{\omega(b) = y(x) = A + B(x + x^2) \ .} \tag{5.88}$$

Now we substitute $x' = x + x^2$ which leads to

$$\boxed{\omega(b) = y(x') = A + Bx' \ ,} \tag{5.89}$$

which can be now fitted with the linear least-square method.

In yet another approach we begin with the fact that the solar rotation parameters A, B, and C from Eq. (5.87) are not orthogonal, which is also the main cause of the cross-talk phenomenon. So, the orthonormalization can be performed using the Gegenbauer polynomials (see, e.g., Morse and Feshbach (1953)) $T^1_{2n}(x)$, with the substitution $x = \sin b$. The procedure is described in detail in Snodgrass (1984), Snodgrass and Howard (1985), and Snodgrass and Ulrich (1990). Instead of Eq. (5.87) the following formula is used:

$$\omega(b) = \bar{A} + \bar{B}\, T^1_2(\sin b) + \bar{C}\, T^1_4(\sin b) \ , \tag{5.90}$$

where the Gegenbauer polynomials are

$$T^1_0(\sin b) = 1 \ , \tag{5.91}$$

$$T^1_2(\sin b) = 5\sin^2 b - 1 \ , \tag{5.92}$$

$$T^1_4(\sin b) = 21\sin^4 b - 14\sin^2 b + 1 \ . \tag{5.93}$$

Finally, we need a transformation between the differential rotation parameters A, B, and C and the other orthogonal set of parameters $\bar{A}$, $\bar{B}$, and $\bar{C}$. Since the Gegenbauer polynomials are linear combinations of the $\sin^2 b$ and $\sin^4 b$, the new parameters are the linear combinations of the old parameters. It can be shown that the following Eqs. are valid for transformation:

$$\boxed{\bar{A} = A + \frac{1}{5}\, B + \frac{3}{35}\, C \ ,} \tag{5.94}$$

$$\boxed{\bar{B} = \frac{1}{5}\, B + \frac{2}{15}\, C \ ,} \tag{5.95}$$

$$\boxed{\bar{C} = \frac{1}{21}\, C \ .} \tag{5.96}$$

So, the orthogonal parameters $\bar{A}$, $\bar{B}$, and $\bar{C}$ can be easily calculated from the parameters A, B, and C obtained by fitting Eq. (5.87). This procedure is widely used in the reduction of data obtained with the Doppler method. Finally, we note that in paper Snodgrass and Ulrich (1990) there is a typing error in one of the transformation formulae, which is corrected here in Eq. (5.96).

5.3.6 Filtering of the Measured Solar Rotation Velocity Data

There are many sources of errors in the measurement of the solar rotation velocity, as we have seen in the previous chapter. So, it is very important to implement an optimal procedure for the rejection of those rotation velocities which are obviously erroneous. It is also important to point out that measurements of the solar rotation reflect the statistical nature of this phenomenon. This means that some spread of observational data is always present, but genuine errors should be excluded as much as possible.

To accomplish this aim, various methods are in use and here we describe briefly one possibility, the so-called two-step velocity filter (Wöhl et al. 2001; Brajša et al. 2001). In Fig. 5.9 one example of solar rotation measurements is presented: rotation velocities in deg/day as a function of latitude in degrees. Data from both solar hemispheres are treated together. Filtering of the erroneous velocity values is performed in two steps:

- All sidereal rotation velocities lower than 8 deg/day and higher than 18 deg/day are excluded. All velocities left after this filter are presented in Fig. 5.9 (both symbols). This is the first filter.

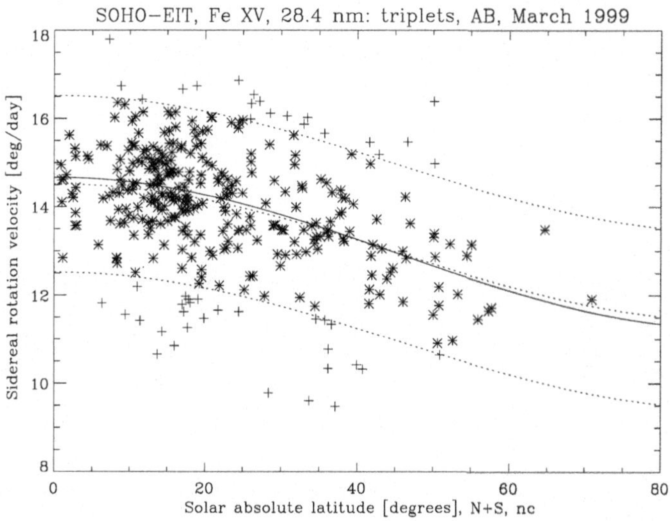

Fig. 5.9 Solar sidereal rotation velocity values obtained by tracing coronal bright points in both hemispheres with the automatic method. The data points excluded by the second filter, as described in the text, are shown by crosses. The fitted differential rotation curve using all data points presented in the Figure, as well as the error ranges differing by 2 deg/day, are denoted by dotted lines. The full line describes the differential rotation curve of the filtered data (stars). Coronal bright points were traced in full-disk solar images taken in March 1999 with the EIT instrument (28.4 nm filter) onboard the SOHO spacecraft. Credit: Wöhl et al. (2001)

- Then, solar rotation parameters from Eq. (5.87) are calculated with the least-square method, and the differential rotation curve is obtained for all data which passed the first filter. Further, a latitude-dependent filter is imposed, excluding all velocity values which differ by 2 deg/day and more from the mean curve. This is the second filter and the data which "survived" it are fitted again with the least-square method to get the final differential rotation profile.

Exercises and Questions

1. Regarding the height correction in the solar rotation determination, show that the following approximative formulae are valid:

$$\sin b_{\mathrm{corr}} \approx \frac{\sin b_{\mathrm{obs}}}{1 + h/R}\,, \tag{5.97}$$

$$\sin CMD_{\mathrm{corr}} \approx \frac{\sin CMD_{\mathrm{obs}} \cos b_{\mathrm{obs}}}{\sqrt{(1 + h/R)^2 - \sin^2 b_{\mathrm{obs}}}}\,, \tag{5.98}$$

where b_{obs} and b_{corr} are the observed and corrected latitudes, CMD_{obs} and CMD_{corr} are the observed and corrected central meridian distances, h is the assumed height of the tracer above the photosphere, and R is the solar radius.

2. Calculate the conversion for different units in which the solar rotation velocity can be expressed. Find the factors by which the deg/day should be multiplied to get the other units, in particular

- deg/day into μrad/s
- deg/day into m/s
- deg/day into nHz.

In which cases the conversion is a constant and when it depends on the latitude?

3. Show that in the linear least-square method the parameters a and b, given in Eqs. (5.77) and (5.78), respectively, can be equivalently expressed as follows:

$$a = \frac{\sum_{i=1}^{N} y_i - b \sum_{i=1}^{N} x_i}{N}\,, \tag{5.99}$$

$$b = \frac{N \sum_{i=1}^{N} (x_i y_i) - \sum_{i=1}^{N} x_i \sum_{i=1}^{N} y_i}{N \sum_{i=1}^{N} x_i^2 - (\sum_{i=1}^{N} x_i)^2}\,. \tag{5.100}$$

4. Show that in the linear least-square method standard errors of the parameters a and b, given in Eqs. (5.79) and (5.80), respectively, can be equivalently expressed as follows:

$$M_a = M_b \sqrt{\frac{1}{N} \sum_{i=1}^{N} x_i^2} \,, \tag{5.101}$$

$$M_b = \sqrt{\frac{1}{N-2}} \sqrt{\frac{N \sum_{i=1}^{N} y_i^2 - (\sum_{i=1}^{N} y_i)^2}{N \sum_{i=1}^{N} x_i^2 - (\sum_{i=1}^{N} x_i)^2} - a^2} \,. \tag{5.102}$$

5. Derive the best-fit parameters a, b, and c for the polynomial of the second degree, see Eq. (5.82), with corresponding standard errors M_a, M_b, and M_c, following a similar procedure as for the linear fit (polynomial of the first degree) described here.

6. Derive Eqs. (5.94), (5.95), and (5.96).

7. Rearrange Eqs. (5.94), (5.95), and (5.96) to get transformation formulae for conversion of the parameters $\bar{A}$, $\bar{B}$, and $\bar{C}$ into the parameters A, B, and C.

References

Anderson, T.W.: The Statistical Analysis of Time Series. Wiley (2011)

Balthasar, H., Lustig, G., Woehl, H., Stark, D.: The solar rotation elements i and omega derived from sunspot groups. Astron. Astrophys. **160**(2), 277–279 (1986)

Balthasar, H., Woehl, H., Stark, D.: The solar rotation elements i and Omega derived from recurrent single sunspots. Astron. Astrophys. **174**(1–2), 359 (1987)

Beck, J.G., Giles, P.: Helioseismic determination of the solar rotation axis. Astrophys. J. Lett. **621**(2), L153–L156 (2005)

Brajša, R., Romštajn, I., Wöhl, H., Benz, A.O., Temmer, M., Roša, D.: Heights of solar tracers observed at 8 mm and an interpretation of their radiation. Astron. Astrophys. **493**(2), 613–621 (2009)

Brajša, R., Wöhl, H., Vršnak, B., Ruždjak, V., Clette, F., Hochedez, J.F., Roša, D.: Height correction in the measurement of solar differential rotation determined by coronal bright points. Astron. Astrophys. **414**, 707–715 (2004)

Brajša, R., Wöhl, H., Vršnak, B., Ruždjak, V., Clette, F., Hochedez, J.F.: Solar differential rotation determined by tracing coronal bright points in SOHO-EIT images. I. Interactive and automatic methods of data reduction. Astron. Astrophys. **374**, 309–315 (2001)

Brajša, R., Wöhl, H., Vršnak, B., Ruždjak, D., Sudar, D., Roša, D., Hržina, D.: Differential rotation of stable recurrent sunspot groups. Solar Phys. **206**(2), 229–241 (2002)

Goldstein, H.: Klassische Mechanik. Akademische Verlagsgesellschaft, Wiesbaden (1983)

Graf, W.: Yearly variation in the synodic rotation period of the Sun. Solar Phys. **37**(1), 257–260 (1974)

Ivezić, Ž, Connolly, A.J., VanderPlas, J.T., Gray, A.: Statistics, Data Mining, and Machine Learning in Astronomy: A Practical Python Guide for the Analysis of Survey Data. Princeton University Press, Princeton and Oxford (2014)

Meeus, J.: Astronomical Algorithms. Willmann-Bell, Richmond (1991)

Morse, P.M., Feshbach, H.: Methods of Theoretical Physics. McGrow-Hill Book Comp, New York (1953)

Press, W.H., Teukolsky, S.A., Vetterling, W.T., Flannery, B.P.: Numerical Recipes in C: The Art of Scientific Computing, 2nd edn. Cambridge University Press, Cambridge (2002)

Roša, D., Brajša, R., Vršnak, B., Wöhl, H.: The relation between the synodic and sidereal rotation period of the Sun. Solar Phys. **159**(2), 393–398 (1995)

Roša, D., Vršnak, B., Božić, H., Brajša, R., Ruždjak, V., Schroll, A., Wöhl, H.: A method to determine the solar synodic rotation rate and the height of tracers. Solar Phys. **179**(2), 237–252 (1998)

Scherrer, P.H., Wilcox, J.M., Svalgaard, L.: The rotation of the Sun - observations at Stanford. Astrophys. J. **241**, 811–819 (1980)

Skokić, I., Brajša, R., Roša, D., Hržina, D., Wöhl, H.: Validity of the relations between the synodic and sidereal rotation velocities of the Sun. Solar Phys. **289**(5), 1471–1476 (2014)

Snodgrass, H.B., Howard, R.: Torsional oscillations of low mode. Solar Phys. **95**(2), 221–228 (1985)

Snodgrass, H.B., Ulrich, R.K.: Rotation of Doppler features in the solar photosphere. Astrophys. J. **351**, 309 (1990)

Snodgrass, H.B.: Separation of large-scale photospheric Doppler patterns. Solar Phys. **94**(1), 13–31 (1984)

Very, F.W.: Heliographic positions. II. Astrophys. J. **6**, 400 (1897)

Vršnak, B., Roša, D., Božić, H., Brajša, R., Ruždjak, V., Schroll, A., Wöhl, H.: Height of tracers and the correction of the measured solar synodic rotation rate: demonstration of the method. Solar Phys. **185**(2), 207–225 (1999)

Waldmeier, M.: Ergebnisse und Probleme der Sonnenforschung, 2nd edn. Akademische Verlagsgesellschaft Geest & Portig K.-G, Leipzig (1955)

Wittmann, A.D.: On the relation between the synodic and sidereal rotation period of the Sun. Solar Phys. **168**(1), 211–213 (1996)

Wöhl, H., Brajša, R., Vršnak, B., Ruždjak, V., Clette, F., Hochedez, J.F.: Determination of the solar rotation tracing EUV bright points with the automatic method. Hvar Obs. Bull. **25**(1), 27–37 (2001)

Chapter 6
A Comparison of Various Observational Results of Solar Differential Rotation

In this chapter we present results of the solar differential rotation measurements obtained by various methods. The main aim is to describe the typical, representative results, to compare them, and try to explain differences and similarities of different observational data. We note that this is certainly not the complete overview of all solar rotation measurements (which would be anyway a rather difficult task), but rather an attempt to give good examples of a variety of observational data with corresponding interpretation, when possible. We begin with the differential rotation of solar plasma, determined below the surface using the helioseismology and in the solar atmosphere with the Doppler method. Further, solar rotation measured with various tracers (sunspots, Hα filaments, tracers observed at mm wavelengths, and coronal bright points) is described. Finally, the results of rotation analyses of solar chromosphere and corona are briefly presented.

6.1 Differential Rotation of Solar Plasma

Before describing various specific results of the solar rotation measurements, a brief reminder about the representation of the solar rotation and the error analysis is given. The results are usually presented as differential rotation curves and with parameters of the solar rotation. So, to make reading this chapter more convenient, the formula of the solar differential rotation is repeated:

$$\boxed{\omega(b) = A + B\sin^2 b + C\sin^4 b \ ,} \tag{6.1}$$

where the parameter A represents the angular rotation rate at the solar equator and the parameters B and C describe the differential character of the solar rotation. The angular rotation velocity is denoted by ω and the latitude by b. We note that all rotation velocities in this chapter are the sidereal ones.

R. Brajša and A. Hanslmeier, *Solar Rotation*, UNITEXT for Physics,
https://doi.org/10.1007/978-981-97-6879-0_6

To check statistical significance of the differences in rotation velocities and/or parameters, a simple criterion is used (see, e.g., Brajša et al. (1999)):

$$\boxed{\Delta\omega \equiv \omega_1 - \omega_2 > n\,M(\omega_1) + n\,M(\omega_2)\,,} \tag{6.2}$$

where ω_1 and ω_2 represent rotation velocities or differential rotation parameters for different measurements, M represents the standard error, and $n = 1, 2, 3$. A difference of measured quantities is statistically significant on the $n\sigma$ level if the criterion (6.2) is fulfilled for n:

- 1σ means a low statistical significance,
- 2σ means a moderate statistical significance, and
- 3σ means a high statistical significance.

Solar rotation determined by plasma motions in the convection zone below the photosphere and in the solar atmosphere can be measured by the helioseismic method and by the Doppler method, respectively. Some representative results of both methods are presented in the next two subsections.

6.1.1 Solar Rotation Below the Photosphere Determined Using the Helioseismic Method

One of the most important results of helioseismology, regarding the solar rotation determination, is reproduced in Fig. 6.1, where the solar rotation velocity is presented as a function of the depth below the photosphere for five different latitudes. The results were obtained by inversion of the GONG data and are adapted from Howe (2009).

Figure 6.1 reveals a lot of information about the complex behavior of plasma rotation in the solar interior. At all latitudes under consideration, $b = 0$ deg, $b = 15$ deg, $b = 30$ deg, $b = 45$ deg, and $b = 60$ deg, immediately below the photosphere rotation velocity increases in comparison with the photospheric values. Going deeper rotation velocity decreases at low latitudes and medium latitudes and remains constant at high latitudes. Further complex patterns can be seen in Fig. 6.1. All five curves converge approximately at the base of or just below the convection zone, where the solar rotation profile becomes less differential and almost uniform (rigid component of solar rotation). Note that a rather different behavior of the solar rotation below the surface was reported earlier from inversion of data obtained with the Michelson Doppler Imager onboard the SOHO satellite (Kosovichev et al. 1997).

In Fig. 6.2 a comparison of the angular velocity of shallow subphotospheric layers (from GONG++ Dopplergrams) and coronal bright points (traced in the SOHO/EIT images) during the declining phase of solar cycle 23 is presented. Two interesting properties can be seen in this figure. First, coronal bright points show systematically higher rotation velocity than plasma at low subphotospheric depths. Second, at low and middle latitudes analyzed here, rotation velocity of plasma below the photosphere increases with increasing depth, in accordance with results presented in Fig. 6.1.

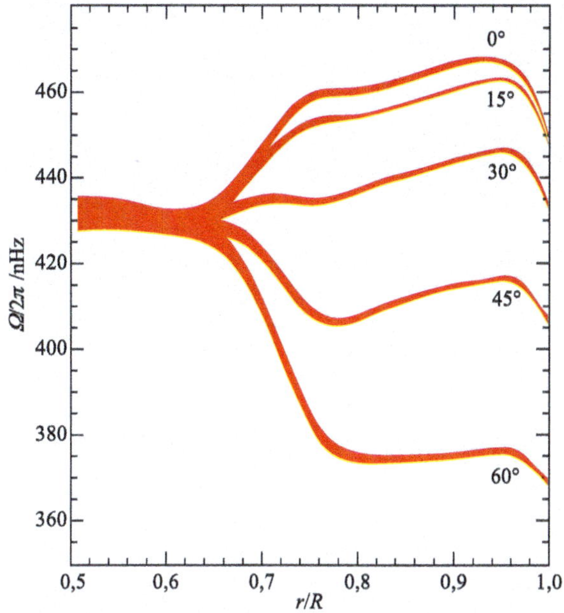

Fig. 6.1 Solar rotation velocity as a function of the distance from the center of the Sun (expressed as a fraction of the solar radius) for five characteristic latitudes, $b = 0$ deg, $b = 15$ deg, $b = 30$ deg, $b = 45$ deg, and $b = 60$ deg. Credit: Wikimedia

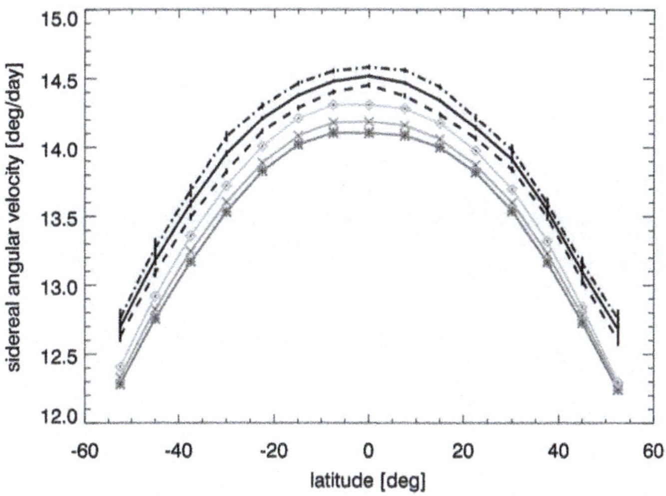

Fig. 6.2 Rotation velocities derived from ring-diagram analysis using GONG++ data at the layers below the solar photosphere at the depths of 3000 km (asterisk), 7000 km (cross), and 15 000 km (diamond) are represented with lines from dark to light gray. Rotation velocities of coronal bright points traced in the SOHO/EIT images are represented with black curves. Positive latitude values refer to the northern solar hemisphere and negative to the southern hemisphere. Credit: Zaatri et al. (2009)

Table 6.1 Parameters of the solar differential rotation from Eq. (6.1) determined with the helioseismic method

Method	Observing period	A [deg d^{-1}]	B [deg d^{-1}]	C [deg d^{-1}]	Source
Helioseismology	1996	14.16	−1.63	−2.52	Schou et al. (1998)
Helioseismology	April 2002	14.04	−1.70	−2.49	Komm et al. (2004)

Examples of results of the helioseismic method about the solar rotation below the solar photosphere are presented through differential rotation parameters in Table 6.1. This representation is important for comparison with the results of other methods.

6.1.2 Rotation of the Solar Atmosphere Determined Spectroscopically with the Doppler Method

Some typical results on the solar rotation determined spectroscopically with the Doppler method are given in Table 6.2. Let us consider the equatorial rotation rate expressed with the differential rotation parameter A. The equatorial rotation velocity is mostly below 14.0 deg/day. Now we compare this value with the corresponding value for the helioseismic method (Table 6.1), which is slightly higher than 14.0 deg/day. Further, both values of the equatorial rotation velocity (plasma at the solar surface and plasma below the solar surface) are significantly lower than the equatorial rotation velocity determined by tracing sunspots, which is about 14.5 deg/day. Results of solar rotation measured by tracing sunspots and sunspot groups will be described in detail in the next subsection.

6.2 Solar Differential Rotation Measured Using Various Tracers

In this section typical results on the solar differential rotation determined by tracing four characteristic, discernible objects in the solar atmosphere are described:

- sunspots and sunspot groups;
- Hα filaments;
- tracers identified in the mm wavelength range;
- coronal bright points.

It is important to note that all of these structures have some relation to the solar magnetic field, which is an important difference to the plasma motion described in the previous section.

Table 6.2 Parameters of the solar differential rotation from Eq. (6.1) determined with the spectroscopic (Doppler) method

Method	Observing period	A [deg d^{-1}]	B [deg d^{-1}]	C [deg d^{-1}]	Source
Doppler	1966–1968	13.76	−1.74	−2.19	Howard and Harvey (1970)
Doppler	1967–1984	14.05	−1.49	−2.61	Snodgrass (1984)
Doppler	1978	13.96 ± 0.10			Perez Garde et al. (1981)
Doppler	1981–1982	13.99 ± 0.06			Kueveler and Woehl (1983)
Doppler	1983–1986	13.92 ± 0.12			Lustig and Woehl (1989)
Doppler	1999	13.01 ± 0.57			Jejčič and Čadež (2005)
Doppler	1999	13.70 ± 1.22			Jejčič and Čadež (2005)

6.2.1 *Solar Differential Rotation Measured by Sunspots and Sunspot Groups*

We begin with two examples of historical measurements of solar rotation using sunspots. Christoph Scheiner measured positions of sunspots in the period 1611–1631. His observing method is illustrated in Fig. 1.6 and an example of his drawing is presented in Fig. 4.14. However, his data became available in digital form, including sunspot positions, only recently (Arlt et al. 2016), which enabled analysis of solar rotation using Scheiner's measurements (Sudar and Brajša 2022). The calculated parameters of solar differential rotation for this curve are presented in Table 6.3, while the solar rotation profile is presented in Fig. 6.3. These results are very important for historical reconstruction of the solar rotation, especially because Scheiner's observations were performed before the Maunder minimum (1645–1715), during which solar activity was significantly reduced.

The second example of historic measurements of solar rotation is presented in Fig. 6.4, where solar rotation velocities of sunspots observed by Rudjer Bošković performed in 1777 are compared with differential rotation profile obtained with modern measurements (Husak et al. 2023). It is noteworthy that both results of these historic solar rotation observations (Scheiner and Bošković) were published very recently, in 2022 (Sudar and Brajša 2022) and in 2023 (Husak et al. 2023), respectively.

Next we present in Fig. 6.5 an example of modern solar rotation determination using sunspot groups from the Debrecen Photoheliographic Data (DPD) catalogue in the period 1974–2016 (Sudar et al. 2017). The differential rotation parameters are

Table 6.3 Values of differential rotation parameters A and B and their standard errors (in deg/day) collected from different sources using sunspots and sunspot groups as tracers. Adapted from Poljančić Beljan et al. (2017), here presented with additional results

Row	Data set	Tracer type	Observing period	A [deg d^{-1}]	B [deg d^{-1}]	Source
1	GPR	a	1874–1976	14.551 ± 0.006	−2.87 ± 0.06	Balthasar et al. (1986)
2	GPR	a	1874–1902	14.63 ± 0.01	−2.70 ± 0.16	Arevalo et al. (1982)
3	GPR	a	1879–1975	14.522 ± 0.005	−2.66 ± 0.04	Javaraiah (2003)
4	GPR	b	1880–1976	14.37 ± 0.01	−2.59 ± 0.16	Brajša et al. (2002)
5	GPR	a	1883–1893	14.63	−2.69	Balthasar and Fangmeier (1988)
6	GPR	a	1940–1968	14.53 ± 0.01	−2.83 ± 0.08	Balthasar and Woehl (1980)
7	GPR	a	1948–1976	14.52	−2.84	Balthasar and Fangmeier (1988)
8	EGR	a	1878–2011	14.49 ± 0.01	−2.64 ± 0.05	Sudar et al. (2014)
9	EGR	a	1976–2002	14.457 ± 0.009	−2.17 ± 0.07	Javaraiah (2003)
10	EGR	c	1874–1996	14.531 ± 0.003	−2.747 ± 0.048	Pulkkinen and Tuominen (1998)
11	CS	c	1853–1893	14.475 ± 0.011	−2.710 ± 0.165	Pulkkinen and Tuominen (1998)
12	Spörer	a	1883–1893	14.50	−2.41	Balthasar and Fangmeier (1988)
13	Abastumani	d	1950–1990	14.73 ± 0.06	−2.07 ± 0.51	Khutsishvili et al. (2002)
14	Mt. Wilson	d	1921–1982	14.522 ± 0.004	−2.84 ± 0.04	Howard et al. (1984)
15	Mt. Wilson	a	1921–1982	14.393 ± 0.010	−2.95 ± 0.09	Howard et al. (1984)
16	DPD	a	1974–2016	14.50 ±0.01	−2.54 ± 0.07	Sudar et al. (2017)
17	KSO	c	1970–1979	14.27 ± 0.02	−1.84 ± 0.12	Lustig (1982)
18	KSO	c	1947–1981	14.38 ± 0.01	−2.57 ± 0.07	Lustig (1983)
19	KSO	a	1948–1976	14.35	−2.73 ±	Balthasar and Fangmeier (1988)
20	KSO	a	1964–2016	14.47 ± 0.01	−2.66 ± 0.10	Poljančić Beljan et al. (2017)
21	KSO	a	1964–2016	14.50 ± 0.01	−2.87 ± 0.12	Poljančić Beljan et al. (2017)
22	Scheiner	a	1611–1631	14.334 ± 0.074	−3.47 ± 0.96	Sudar and Brajša (2022)

Notes

Data set:
GPR—Greenwich Photoheliographic Results
EGR—extended Greenwich Results
CS—Carrington & Spörer data
DPD—Debrecen Photoheliographic Results
KSO—Kanzelhöhe Observatory for Solar and Environmental Research data

Tracer type:
a—sunspot groups
b—stable recurrent sunspot groups
c—sunspots and sunspot groups
d—sunspots

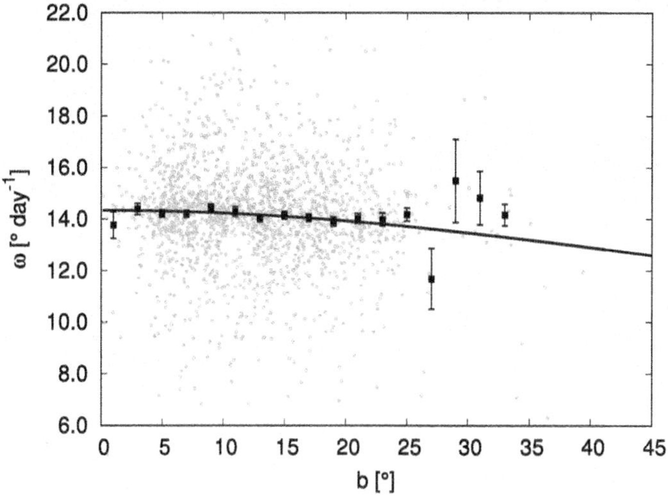

Fig. 6.3 Rotation velocities of sunspot groups calculated from position measurements of sunspots performed by Ch. Scheiner in the period 1611–1631. Individual velocities are shown as gray dots and the best-fit curve is shown with a solid line. Average rotational velocities in latitude bins of 2 degrees are shown as black squares with standard deviations indicated as vertical error bars. The differential rotation parameters are given in Table 6.3, row 22. Credit: Sudar and Brajša (2022)

given in Table 6.3. When all sunspot types are included in the analysis and several solar cycles are covered, as in this case, the sidereal rotation velocity at the solar equator is about 14.5 deg/day. Note that this value is determined with a rather high reliability based on very small measurement errors at low and middle latitudes. Now we compare this result with the Doppler measurements (see the previous subsection) and it can be seen that sunspots rotate systematically faster than the surrounding plasma. On the other hand, if a subgroup of only stable recurrent sunspot groups is considered (Fig. 6.6), it can be seen that a systematically lower differential rotation profile is measured. Such sunspot groups consist of long-living sunspots which survived several solar rotations. This behavior, also seen in Table 6.3, can be explained with the drag of ambient plasma, which slows down sunspots if they live long enough.

Differential rotation profiles determined by tracing sunspots and sunspot groups using Kanzelhöhe data set in various temporal subintervals are presented in Fig. 6.7. A comparison of resulting differential rotation curves obtained using data from various observatories and time intervals is given in Fig. 6.8. In these two figures selected differential rotation profiles are presented, based on parameters listed in Table 6.3. The spread of the family of differential rotation curves is caused by various factors, including the selection process of tracers, the reduction procedure, the time span, etc.

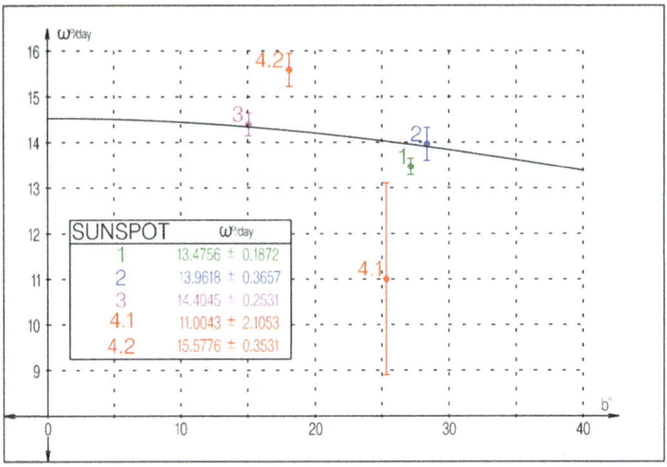

Fig. 6.4 Rotation velocities of five sunspots observed by R. Bošković during several days in September 1777 are presented with small circles with corresponding standard deviations presented with error bars. Differential rotation profile of sunspot groups from the period 1874–1976 (Ruždjak et al. 2017), defined with the rotation parameters $A = (14.528 \pm 0.006)$ deg/day and $B = (-2.77 \pm 0.05)$ deg/day, is given for comparison. Credit: Husak et al. (2023)

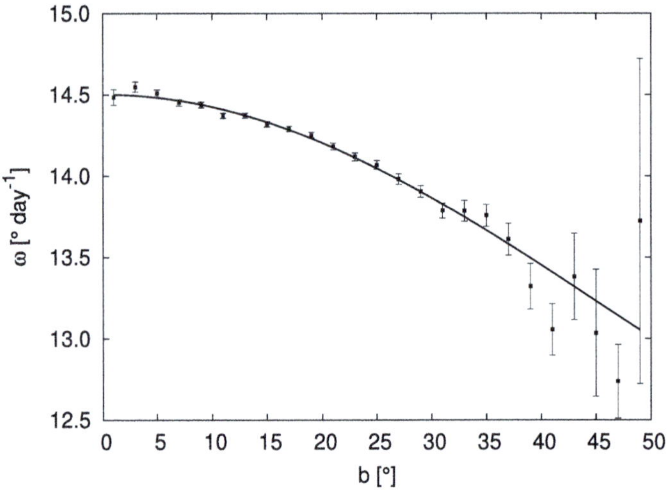

Fig. 6.5 Solar differential rotation determined by tracing sunspot groups in the period 1974–2016. The Debrecen Photoheliographic Data (DPD) catalogue is used. Differential rotation profile is presented with the solid line. Filled squares with error bars are averaged values for the 2 deg latitude bins. The differential rotation parameters are given in Table 6.3, row 16. Credit: Sudar et al. (2017)

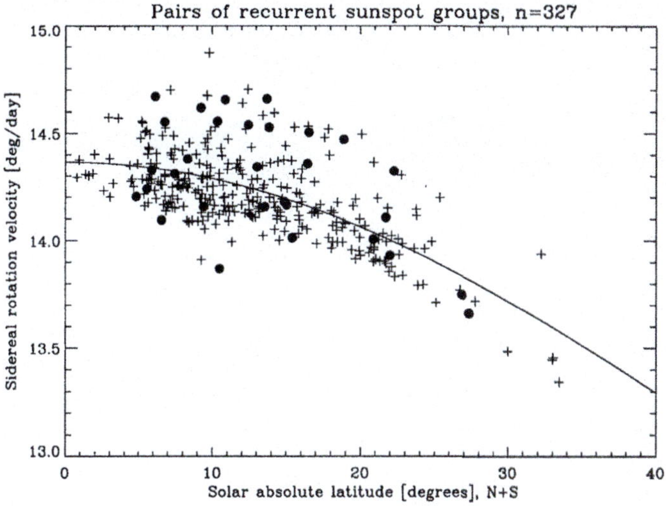

Fig. 6.6 Sidereal rotation velocities obtained measuring the time between two successive passages across the central meridian (the period method) for 327 stable recurrent sunspot groups from the Greenwich data set (DPR). Filled circles represent velocities of growing groups and crosses represent velocities of decaying groups. Both solar hemispheres are treated together and the full line gives the differential rotation profile for all data points. The differential rotation parameters are given in Table 6.3, row 4. Credit: Brajša et al. (2002)

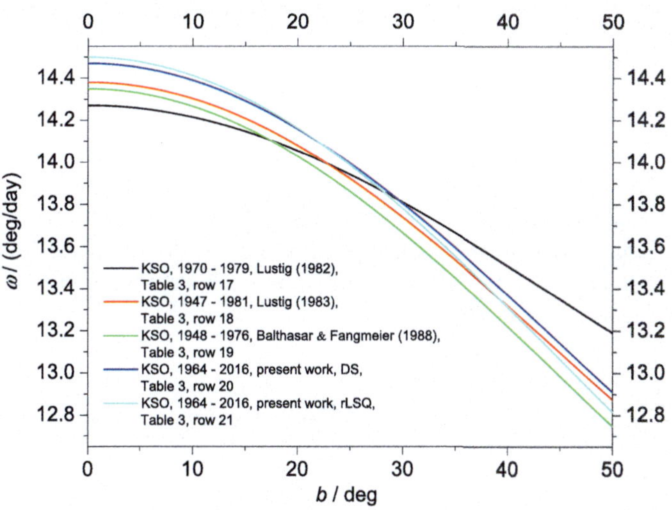

Fig. 6.7 Differential rotation profile calculated by various authors and for various time spans using sunspot data obtained at Kanzelhöhe Observatory for Solar and Environmental Research (KSO). The rows in the legend refer to Table 6.3. Credit: Poljančić Beljan et al. (2017)

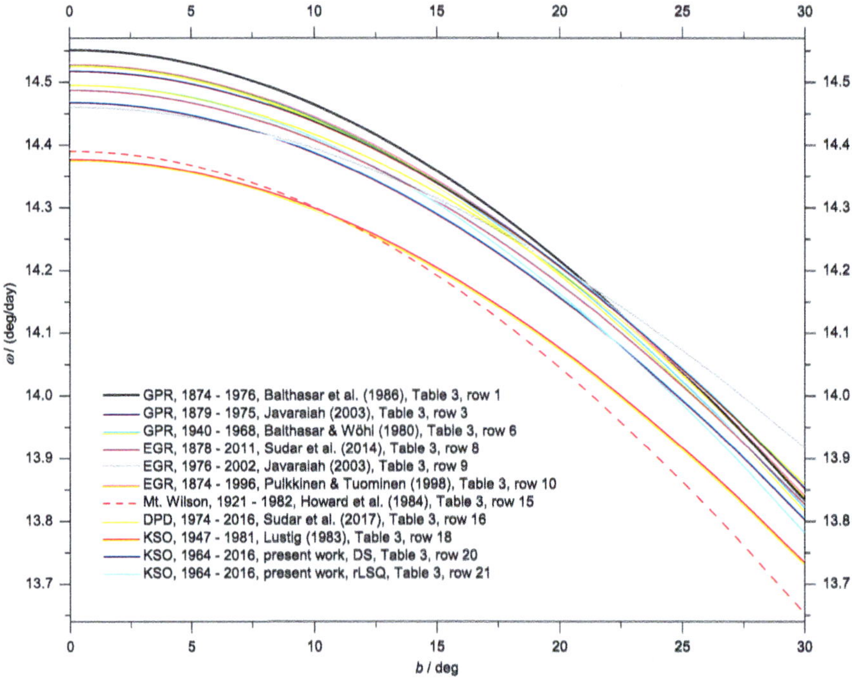

Fig. 6.8 Differential rotation profile calculated by various authors and for various time spans using sunspot data from different data sets, as indicated in the legend. The rows in the legend refer to Table 6.3. Credit: Poljančić Beljan et al. (2017)

Finally, we should address the question of why sunspots exhibit larger rotation velocity than plasma in which they are embedded. This can be explained with the fact that rotation velocity increases with depth in the convection zone, as revealed by helioseismic measurements. Sunspots are magnetic flux tubes anchored deep in the convection zone and they bring information of the larger rotation velocity from the deeper layers to the surface of the Sun.

6.2.2 *Solar Differential Rotation Measured by Hα Filaments*

Solar differential rotation parameters obtained by tracing Hα filaments from some selected studies are presented in Table 6.4. In some cases the approximative height correction procedure was applied, as indicated in the table, implying that those results are in principle transformed into the photospheric level. It is interesting to note that the solar rotation parameter A, describing the rotation velocity at the solar equator, was found to be just below 14.5 deg/deg, very close to the same quantity for the sunspots. However, the slope of the differential rotation curve is different from sunspots. For

Table 6.4 Parameters of the solar differential rotation from Eq. (6.1) determined by tracing Hα filaments. Application of the height correction procedure is denoted with "h.c.". The Coimbra data refer to filaments from the Ca II K3 filtergrams

Data set	Observing period	A [deg d^{-1}]	B [deg d^{-1}]	C [deg d^{-1}]	Source
Meudon	1919–1929	14.48	−2.16		d'Azambuja and d'Azambuja (1948)
Big Bear, with h.c.	1972–1973	14.45	−1.43		Adams and Tang (1977)
Big Bear and Kanzelhöhe, with h.c.	1972–1987	14.45 ± 0.15	−0.11 ± 0.90	−3.69 ± 0.90	Brajša et al. (1991)
Coimbra, with h.c.	1929–1941	14.91 ± 0.26	−3.51 ± 0.68		Wan and Li (2022)

comparison the recently published results obtained by tracing filaments in the Ca II K3 filtergrams obtained at the Geophysical and Astronomical Observatory of the University of Coimbra (OGUAC) are presented in the last row in Table 6.4 (Wan and Li 2022). It can be seen that these data (Ca II K3 filaments) indicate a higher equatorial rotation velocity and a steeper differential rotation as compared with the Hα filaments.

6.2.3 *Solar Differential Rotation Measured by Structures Identified in the mm Wavelength Range*

Full-disk solar images recorded in the mm wavelength range using radio telescopes are very useful data sources for solar rotation determination. Various identified objects can be traced in a series of images for measuring solar rotation velocity. A relatively short observing wavelength enables good spatial resolution in comparison with most radio observations at cm and longer wavelength. Here we present two examples of solar rotation determination using data obtained at mm wavelengths.

In Fig. 6.9 differential rotation profiles of tracers followed in full-disk solar images obtained at a wavelength of 8 mm are presented. Daily full-disk solar maps were taken in the years 1979–1982 and 1987–1991 with the 14-m antenna of the Metsähovi Radio Observatory, Helsinki University of Technology, Finland (see Brajša et al. (2009) and references therein). Two types of objects were traced: the so-called high temperature regions (HTRs) and low temperature regions (LTRs). In radio astronomy the brightness temperature is a measure of the radiation intensity. The HTRs have the brightness temperature higher than the quiet Sun level and are directly linked to active regions. Their source radiation is located above active regions. The LTRs have the

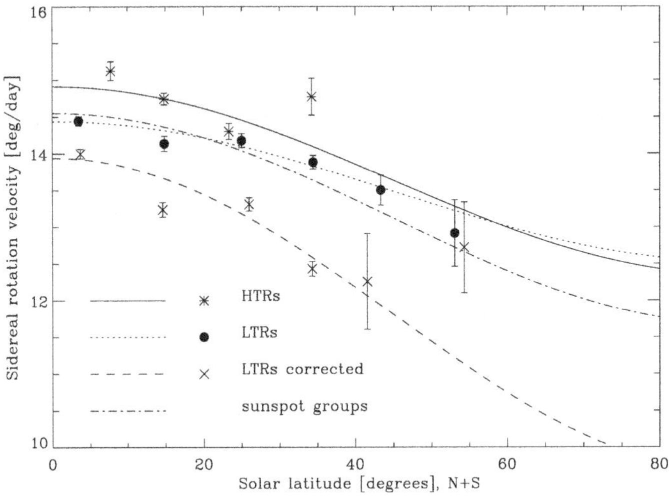

Fig. 6.9 The mean values of sidereal rotation velocities of high temperature regions (HTRs) and low temperature regions (LTRs) observed at the wavelength of 8 mm using the 14-m radio dish of the Metsähovi Radio Observatory, Finland. For the LTRs, results with and without the height correction are presented, as indicated in the legend. The averaging is performed in 10 deg latitude bins and the error bars indicate the standard errors of the means in each bin. Both solar hemispheres are treated together and the lines represent least-square fits to Eq. (6.1) with $C = 0$, using all filtered data points for each case. For comparison, the differential rotation profile obtained by tracing sunspot groups from the Greenwich data set Balthasar et al. (1986) is also given. Credit: Brajša et al. (2009)

brightness temperatures lower than the quiet Sun level and are related to Hα filaments and inversion lines of the large-scale magnetic field. In Fig. 6.9 differential rotation curves are presented together with the mean values of individual rotation velocities for the 10 deg latitude bins. For LTRs the height correction was also performed and it can be seen how the velocity values and corresponding latitudes changed. As a result of this procedure the solar rotation curve was shifted to lower values and became more differential. Finally, the differential rotation profile obtained by tracing sunspot groups is also given for comparison. Differential rotation profile of HTRs is similar to the curve of sunspot groups, but is shifted systematically to higher values. One would expect that HTRs and sunspot groups (manifestations of active regions) will expose the same rotational behavior, but the systematic difference can be explained by the fact that the height correction of HTRs was not performed, due to lack of sufficient data.

The second example, presented in Fig. 6.10, refers to results obtained by tracing identified structures in the ALMA full-disk solar images obtained with the single-dish measurements at the wavelength of 1.3 mm. ALMA is the acronym for the Atacama Large Millimeter/Submillimeter Array, an international radio observatory located in the Atacama desert in Chile. A series of solar images taken on April 13, 2017 was analyzed and an automatic method for measuring the differential rotation profile of the Sun from short time span data, based on tracking features with an

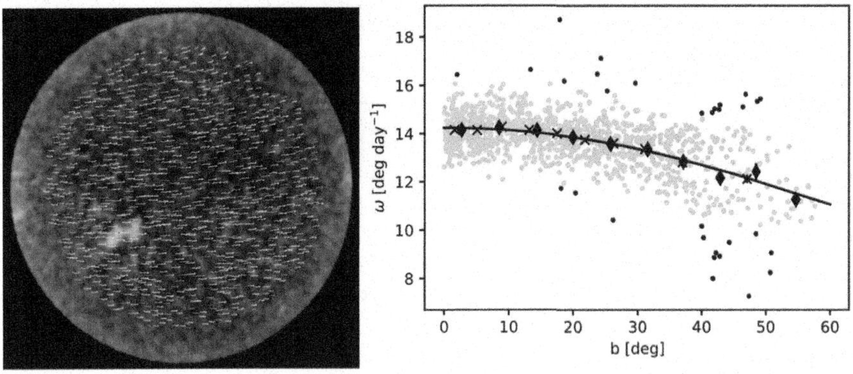

Fig. 6.10 Solar structures observed and traced in the ALMA full-disk solar images taken at a wavelength of 1.3 mm using a single-dish radio antenna. In the left panel traces of identified objects are denoted by colored lines. In the right panel gray circles are individual measurements, black circles are discarded measurements (which are filtered out), black crosses indicate bins with equal number of elements, black diamonds are bins with equal widths, and the black solid line is the final differential rotation fitted profile. Credit Skokić et al. (2022)

optical flow algorithm and robust iterative fitting, was developed. Now we compare some characteristics of solar observations using the Metsähovi and ALMA radio telescopes. The ALMA measurements, even in the single-dish mode (ALMA has also the interferometric capability), have better spatial resolution due to the shorter observing wavelengths. At these wavelengths the source region in the solar atmosphere is in the chromosphere, i.e., at a much lower height than the Metsähovi HTRs and LTRs. Finally, we compare the solar rotation profiles obtained with the Metsähovi and ALMA measurements; see Figs. 6.9 and 6.10. It is noteworthy that the height corrected Metsähovi LTR curve (Fig. 6.9) is very close to the ALMA "raw" (height uncorrected) curve (Fig. 6.10). This result is consistent with the fact that the average height of the Metsähovi LTRs is about 45,000 km, while the typical height of the ALMA sources at the given wavelength is much lower, on the order of a couple of thousands km. We also note that the analysis of the solar rotation using the ALMA data is work in progress and the results presented here are by no means the final result, but rather a preliminary one, at the level of proof of concept for further investigations.

6.2.4 Solar Differential Rotation Measured by Coronal Bright Points

Coronal bright points are very appropriate tracers for the solar rotation measurements, as we have already seen in Chap. 4. These objects are traced in images obtained with the ultraviolet and X-ray telescopes placed outside the Earth's atmosphere. Here we list a couple of representative examples of such research performed with instruments on satellites and space stations:

- OSO-4, Mg X spectroheliograms, 1967, Dupree and Henze (1972);
- Skylab, soft X-rays, 1973, Golub et al. (1974);
- Yohkoh/SXT, soft X-rays, 1992–2001, Kariyappa (2008);
- Yohkoh/SXT, soft X-rays, 1993, Brajsa et al. (2001);
- Yohkoh/SXT, soft X-rays, 1994–1997, Hara (2009);
- SOHO/EIT, 19.5 nm, 1996, Karachik et al. (2006);
- SOHO/EIT, 28.4 nm, 1998–2006, Brajša et al. (2004), Wöhl et al. (2010);
- SOHO/EIT, 19.5 nm, 2005, Karachik et al. (2006);
- Hinode/XRT, soft X-rays, 2007, Kariyappa (2008);
- SDO/AIA, 2011, 19.3 nm, Sudar et al. (2015, 2016).

From this list we clearly see how valuable are the space missions for solar physics research, including the special subtopic of solar rotation investigation.

We now discuss in more detail solar differential rotation determined by tracing coronal bright points using data from two satellite missions, SOHO and SDO.

In Fig. 6.11 a differential rotation curve obtained by tracing coronal bright points in SOHO/EIT images during 1998–1999 is presented (Brajša et al. 2004). The 28.4

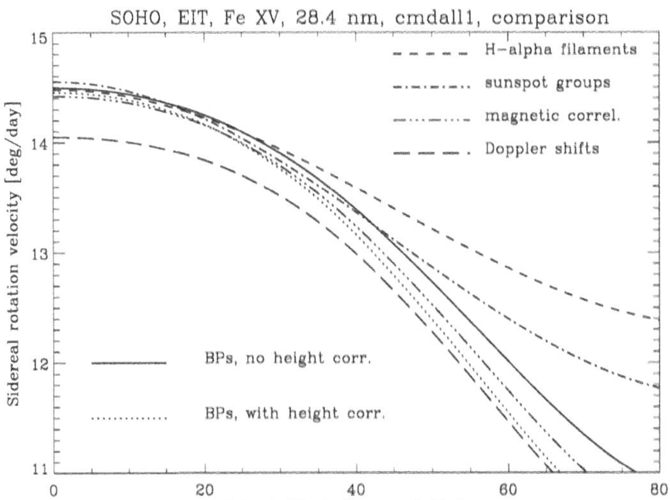

Fig. 6.11 A comparison of the solar differential rotation curves determined by tracing coronal bright points in the SOHO/EIT images Brajša et al. (2004) and other results obtained with various tracers and methods. In the case of coronal bright points (BPs) the interactive method was used, and the results with and without the height correction are presented, as indicated in the legend (the case with $B = C$, see Eq. (6.1)). The solar rotation profiles for Hα filaments d'Azambuja and d'Azambuja (1948), sunspot groups Balthasar et al. (1986), correlations of small magnetic elements Komm et al. (1993), and Doppler shifts Snodgrass (1984) are given for comparison. Adapted from Brajša et al. (2004)

nm channel was used with a cadence of 6 hours (4 images per day). The measurements were performed using the interactive method (Brajša et al. 2001, 2002) based on a visual identification of coronal bright point tracers in images with the spatial resolution of 2.629"/pixel. Results with and without the height correction are presented and compared with the differential rotation profiles of Hα filaments, sunspot groups, correlated small magnetic elements, and plasma measured with the Doppler method. The results of the height correction procedure are clearly seen in a systematic shift of the differential rotation curve to lower values. A close correspondence between the differential rotation profiles of coronal bright points and small photospheric magnetic features was found. This can be interpreted by the fact that coronal bright points lie over the small bipolar magnetic structures in the photosphere, representing the so-called ephemeral magnetic regions, where magnetic flux is either emerging or canceling.

Since the interactive method of tracing coronal bright points is rather time-consuming, the automatic method was developed (Brajša et al. 2001, 2002). The automatic method is significantly faster in its performance and a much larger data set could have been reduced in comparison with the interactive method. So, a precise differential rotation formula was obtained with the automatic method using data from 93 months in the period 1998–2006 (Wöhl et al. 2010). The following solar rotation parameters according to Eq. (6.1) were calculated (in deg/day):

$$\omega(b) = 14.499(\pm 0.006) - 2.54(\pm 0.06)\sin^2 b - 0.77(\pm 0.09)\sin^4 b \ . \tag{6.3}$$

A next huge leap forward towards a high precision determination of the solar rotation profile was enabled using SDO/AIA data (Sudar et al. 2015, 2016). This improvement was made possible due to better instrument capabilities compared to the previous similar space missions. The analyzed full-disk solar images were taken in the 19.3 AIA channel of the SDO satellite with a spatial resolution of about 0.6"/pixel and with a cadence of 10 minutes between successive images. As a proof of concept data from only two days (January 1–2, 2011) were analyzed first (Sudar et al. 2015). In Fig. 6.12 distribution of velocity vectors of traced coronal bright points is presented. A very good latitudinal and longitudinal coverage of data can be seen. It is also important to note how the observation of the solar differential rotation is a statistical phenomenon with a mean, organized flow and individual, fluctuating proper motions of tracers. Further, we present in Fig. 6.13 differential rotation curves for the 2-day data set (January 1–2, 2011), together with the average velocity values in the 5 deg latitude bins, for more than 10,000 individual data points. In this figure it can be seen how different fitting procedures can yield different rotation profiles for the same analyzed data set (see also Eq. (6.1) and legend in the figure). Finally, the described procedure was extended to the data taken in the first five months of 2011 (Sudar et al. 2016). More than 80,000 velocity measurements were analyzed and the

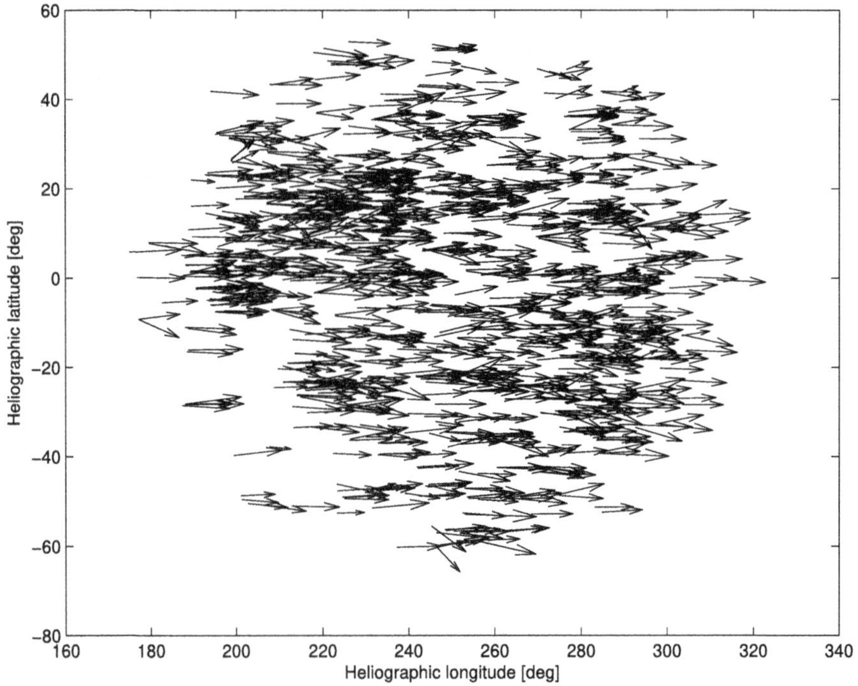

Fig. 6.12 Distribution of coronal bright points in heliographic coordinates (the latitude and the longitude) with arrows showing the direction and magnitude of velocity vectors. The used data are from SDO/AIA. Credit: Sudar et al. (2015)

result is given in Fig. 6.14. The following high precision solar differential rotation parameters according to Eq. (6.1) were calculated (in deg/day):

$$\omega(b) = 14.4060(\pm 0.0051) - 1.662(\pm 0.050)\sin^2 b - 2.742(\pm 0.081)\sin^4 b\ . \tag{6.4}$$

6.3 Rotation of the Higher Layers in the Solar Atmosphere

Measurements of the rotation velocity in the layers of the solar atmosphere higher than the photospheric layer are more difficult and complicated than, e.g., tracing sunspots. Also, the results are more ambiguous in comparison with those regarding rotation of the solar photosphere. We note that comprehensive overviews about this topic can be found in the following reviews: Howard (1975, 1984), Schroeter (1985), Antonucci et al. (2020), and references therein.

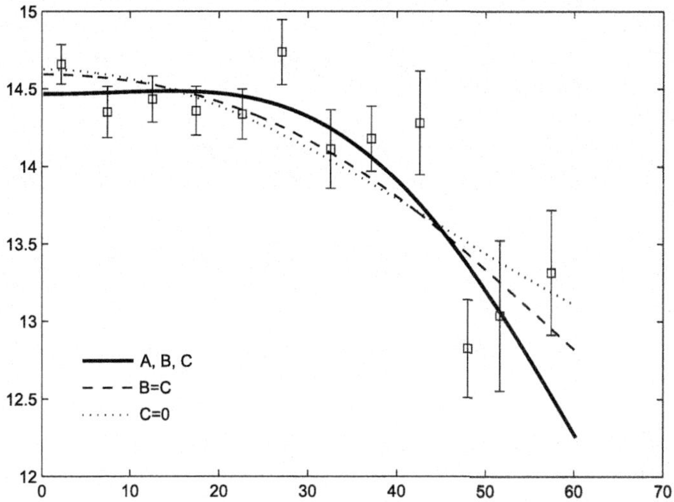

Fig. 6.13 A comparison of three different fitting procedures for the solar differential rotation profile of coronal bright points traced in the SDO/AIA images. See the legend and Eq. (6.1). On the x-axis is the latitude in deg and on the y-axis the sidereal rotation velocity in deg/day. The average values of rotation velocities in the 5 deg latitude bins are also shown with their respective errors. Results for two observing days are presented; for details see text. Credit: Sudar et al. (2015)

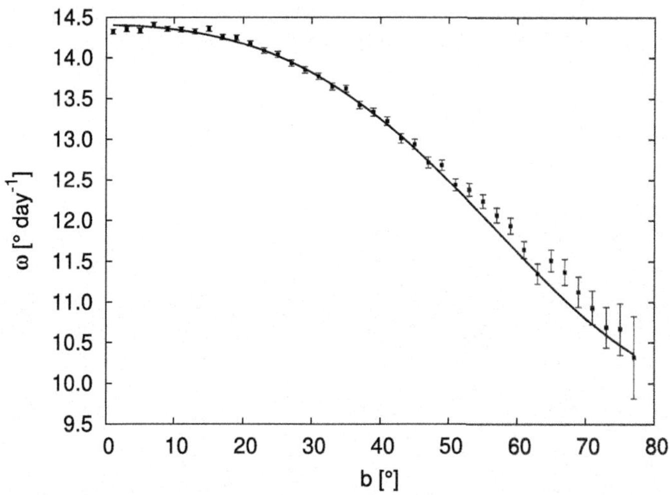

Fig. 6.14 The mean values of the solar rotation velocity ω in 2 deg bins of latitude b denoted with black squares and corresponding error bars for coronal bright points traced in the SDO/AIA images. The best-fit rotation profile is denoted with the solid black line. Results for about 5 months of observations are presented; for details see text. Credit: Sudar et al. (2016)

Although there are some inconsistencies in quite a few observations, there is a general consensus that the rotation velocity increases with height in the solar atmosphere and becomes less differential (i.e., showing a rigid rotation component). Additionally, as a rule of thumb we can say that the larger the structure and the longer its lifetime, this rigid component is more pronounced.

It should be also stressed that in the previous subsections we already treated solar rotation using tracers which belong to the chromosphere and corona, such as Hα filaments, radio sources in the mm wavelength range, and coronal bright points. However, in many of those cases the higher and more rigid observed rotation velocity can be attributed to the projection effects due to the altitude of the tracers and by an application of the height correction the final differential rotation profile is recalibrated to the photospheric level.

In the next subsections we briefly present the most important results about rotation velocity of the solar chromosphere, corona, and coronal holes.

6.3.1 *Rotation of the Solar Chromosphere*

The rotation velocity of the solar chromosphere is faster by several percent than the rotation velocity of the photosphere and the differential character is less pronounced, consistent with a rigid rotational component at the middle and high latitudes. This might be an indication that in the chromosphere the rotation of the magnetic field dominates over the plasma motions, unlike in the photosphere (Howard 1984).

Using the Doppler method systematically higher rotation velocities were measured in the solar chromosphere (Livingston 1969; Livingston et al. 1979) compared to the Doppler measurements at the photospheric level. Moreover, tracing of Ca^{+} mottles also confirmed a higher rotation velocity of the solar chromosphere (Schroeter and Woehl 1976) as compared to the Doppler measurements in the photosphere. However, a cross-correlation analysis of the EUV spectroheliograms revealed a lower (slower) differential rotation profile of the solar chromosphere and corona (Dupree and Henze 1972; Henze and Dupree 1973), similar to, or even below, the Doppler values for the photosphere, but with a rigid rotation component at medium latitudes. This discrepancy might be a result of the used method and is further discussed in Howard (1984).

Some recent studies enabled an investigation of the long-term behavior of the chromospheric rotation. Here we list the most important data sets and their analyses:

- Mount Wilson, Ca II K spectroheliograms, 1915–1985, Bertello et al. (2020), Wan et al. (2023);
- NSO/Kitt Peak, He I 1083.0 nm spectroheliograms, 2004–2013, Li and Xu (2024).

In those papers possible cycle-related variations and north–south asymmetry of the chromospheric rotation were investigated, as well as possible changes in rotation velocity between the active and quiet solar chromosphere. These subtopics will be discussed in more detail in Chap. 8.

6.3.2 Rotation of the Solar Corona

Rotation of the solar corona has been studied using several methods (autocorrelation, stackplots, synoptic charts) and data from ground-based coronameters and instruments on satellites (e.g., Hansen et al. (1969), Kozuka et al. (1994), Weber et al. (1999), Altrock (2003), and Rybak (1994, 2000)). While a general agreement with other methods and data in the equatorial regions was found, a strong rigid rotation behavior was revealed at medium and high latitudes. This means that the differential character of the solar rotation is still present, but is significantly reduced in the solar corona. Also, temporal variations of the coronal differential rotation profile were found, mostly related to the solar cycle phase of observation.

In a recent paper (Singh et al. 2021) rotation of the solar corona was studied using the period analysis of the disk-integrated radio flux recorded at 15 different frequencies in the MHz range. Since various radio frequencies correspond to various heights in the solar atmosphere, this investigation enabled an investigation of the solar rotation as a function of height in the solar corona.

Rotation properties of the solar corona have an important influence on the heliospheric physics. According to the present concept the sector-structure of the interplanetary magnetic field and of the solar wind are strongly correlated with the large-scale, long-lived coronal magnetic structures. These structures expose a strong rigid rotation component (Schroeter 1985; Wang et al. 1988; Nash 1991a, b).

6.3.3 Rotation of Coronal Holes

A subtopic of special interest is an investigation of solar rotation using coronal holes. An important milestone in this research was enabled by using the Skylab and other satellite data (soft X-ray and EUV) in the 1970s (Golub and Pasachoff 2009). A clear rigid rotation of coronal holes was found with a small differential component. This result was later confirmed with other data and methods (e.g., Obridko and Shelting (1989), Navarro-Peralta and Sanchez-Ibarra (1994), Insley et al. (1995), and Ikhsanov and Ivanov (1999)). It is important to note that a pronounced variability of the (small) differential rotation character of coronal holes was found, dependent on the solar cycle phase (activity minimum, maximum, or intermediate phase) and a coronal hole type (polar, equatorial, etc.). The rigid rotation of coronal holes can be understood using a current-free model of the coronal magnetic field, including differential rotation, supergranular diffusion, and systematic meridional motion (Nash et al. 1988). The model was successfully applied to explain the behavior of the so-called "coronal hole 1" observed by Skylab (Wang and Sheeley 1993).

An interesting example of the rigid component of the coronal hole rotation provided by indirect measurements is given in Fig. 6.15. An extensive set of coronal bright points rotation velocities based on the SDO/AIA data, discussed earlier, is here divided into two subsets: the tracers located within coronal holes and the ones

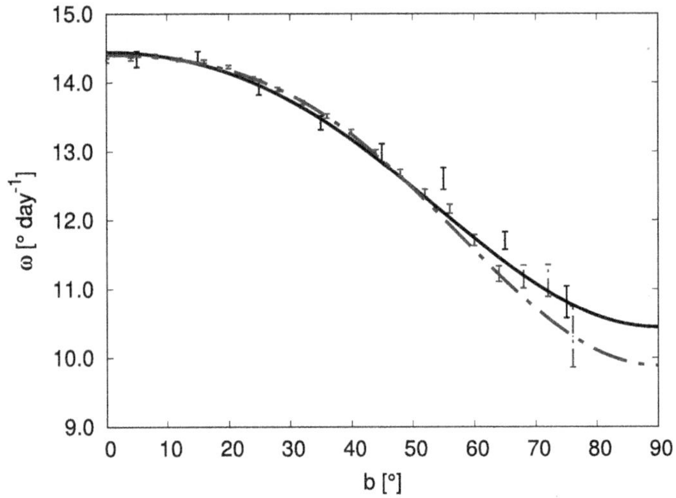

Fig. 6.15 Similar to Fig. 6.14, but analyzed separately for two data subsets: coronal bright points in coronal holes (solid black line) and coronal bright points outside coronal holes (gray dot-dashed line). The latitude bin averaged values with error bars are also presented for both data subsets. Credit: Sudar et al. (2024)

located outside of coronal holes. As can be clearly seen in Fig. 6.15 coronal bright points observed within the coronal hole borders show a rigid component in their rotation rate at high latitudes, obviously bringing information from the coronal hole rotation pattern (Sudar et al. 2024).

Exercises and Questions

1. Describe rotation of solar plasma in the radiative zone, in the convection zone, and at the solar surface.

2. Calculate the differential rotation parameters A, B, and C in Table 6.2, second row, from the original paper Snodgrass (1984). Take into account different units and transformation from differential rotation formula using Gegenbauer polynomials and usual formula Eq. (6.1), as described at the end of the previous chapter.

3. Why is the equatorial rotation velocity determined by tracing sunspots systematically higher than the corresponding value from Doppler measurements?

4. Calculate the beam size on the solar radio maps obtained by scanning the whole solar disk at a wavelength of 8 mm with the 14-m dish antenna of the Metsähovi Radio Observatory. The beam size is a measure of the spatial resolution in radio astronomy. Compare the calculated value with a combination of the wavelength and antenna size using the single-dish observations at the ALMA observatory.

5. What are the main satellite/space missions used for solar research in the ultraviolet and X-ray parts of the spectrum? Which space agencies were involved in preparing those space missions?

6. Which other tracers have the most similar differential rotation profile as the coronal bright points?

7. Describe the behavior of solar differential rotation in the chromosphere and corona.

References

Adams, W.M., Tang, F.: Differential rotation of short-lived solar filaments. Solar Phys. **55**(2), 499–504 (1977)

Altrock, R.C.: A study of the rotation of the solar corona. Solar Phys. **213**(1), 23–37 (2003)

Antonucci, E., Harra, L., Susino, R., Telloni, D.: Observations of the solar corona from space. Space Sci. Rev. **216**(8), 117 (2020)

Arevalo, M.J., Gomez, R., Vazquez, M., Balthasar, H., Woehl, H.: Differential rotation and meridional motions of sunspots from 1874 to 1902. Astron. Astrophys. **111**(2), 266–271 (1982)

Arlt, R., Senthamizh Pavai, V., Schmiel, C., Spada, F.: Sunspot positions, areas, and group tilt angles for 1611-1631 from observations by Christoph Scheiner. Astron. Astrophys. **595**, A104 (2016)

Balthasar, H., Fangmeier, E.: Comparison of the differential rotation laws and meridional motions determined from sunspot positions taken from the Greenwich Photoheliographic Results, the drawings of G. Spoerer, and the Kanzelhoehe data. Astron. Astrophys. **203**(2), 381–384 (1988)

Balthasar, H., Vazquez, M., Woehl, H.: Differential rotation of sunspot groups in the period from 1874 through 1976 and changes of the rotation velocity within the solar cycle. Astron. Astrophys. **155**(1), 87–98 (1986)

Balthasar, H., Woehl, H.: Differential rotation and meridional motions of sunspots in the years 1940–1968. Astron. Astrophys. **92**(1–2), 111–116 (1980)

Bertello, L., Pevtsov, A.A., Ulrich, R.K.: 70 years of chromospheric solar activity and dynamics. Astrophys. J. **897**(2), 181 (2020)

Brajša, R., Romštajn, I., Wöhl, H., Benz, A.O., Temmer, M., Roša, D.: Heights of solar tracers observed at 8 mm and an interpretation of their radiation. Astron. Astrophys. **493**(2), 613–621 (2009)

Brajša, R., Ruždjak, V., Vršnak, B., Wöhl, H., Pohjolainen, S., Urpo, S.: An estimate of microwave low-brightness-temperature regions' heights obtained measuring their rotation velocity. Solar Phys. **184**(2), 281–296 (1999)

Brajsa, R., Vrsnak, B., Ruzdjak, V., Rosa, D., Hrzina, D., Wöhl, H., Clette, F., Hochedez, J.F.: An analysis of the solar rotation velocity by tracing coronal features. In: Brekke, P., Fleck, B., Gurman, J.B. (eds.) Recent Insights into the Physics of the Sun and Heliosphere: Highlights from SOHO and Other Space Missions, vol. 203, p. 377 (2001)

Brajša, R., Vršnak, B., Ruždjak, V., Schroll, A., Pohjolainen, S., Urpo, S., Teräsranta, H.: Solar differential rotation determined by polar crown filaments. Solar Phys. **133**(2), 195–203 (1991)

Brajša, R., Wöhl, H., Vršnak, B., Ruždjak, V., Clette, F., Hochedez, J.F., Roša, D.: Height correction in the measurement of solar differential rotation determined by coronal bright points. Astron. Astrophys. **414**, 707–715 (2004)

Brajša, R., Wöhl, H., Vršnak, B., Ruždjak, V., Clette, F., Hochedez, J.F.: Solar differential rotation determined by tracing coronal bright points in SOHO-EIT images. II. Results for 1998/99 obtained with interactive and automatic methods. Astron. Astrophys. **392**, 329–334 (2002)

Brajša, R., Wöhl, H., Vršnak, B., Ruždjak, V., Clette, F., Hochedez, J.F.: Solar differential rotation determined by tracing coronal bright points in SOHO-EIT images. I. Interactive and automatic methods of data reduction. Astron. Astrophys. **374**, 309–315 (2001)

Brajša, R., Wöhl, H., Vršnak, B., Ruždjak, D., Sudar, D., Roša, D., Hržina, D.: Differential rotation of stable recurrent sunspot groups. Solar Phys. **206**(2), 229–241 (2002)

d'Azambuja, L., d'Azambuja, M.: Ann. Obs. Paris, Meudon **6**(7) (1948)

Dupree, A.K., Jr. Henze, W.: Solar rotation as determined from OSO-4 EUV spectroheliograms. Solar Phys. **27**(2), 271–279 (1972)

Golub, L., Krieger, A.S., Silk, J.K., Timothy, A.F., Vaiana, G.S.: Solar x-ray bright points. Astrophys. J. Lett. **189**, L93 (1974)

Golub, L., Pasachoff, J.M.: The Solar Corona, 2nd edn. Cambridge University Press, Cambridge (2009)

Hansen, R.T., Hansen, S.F., Loomis, H.G.: Differential rotation of the solar electron corona. Solar Phys. **10**(1), 135–149 (1969)

Hara, H.: Differential rotation rate of x-ray bright points and source region of their magnetic fields. Astrophys. J. **697**(2), 980–984 (2009)

Henze, W., Dupree, A.K.: Solar rotation in the chromosphere and corona. Solar Phys. **33**(2), 425–429 (1973)

Howard, R., Gilman, P.I., Gilman, P.A.: Rotation of the sun measured from Mount Wilson white-light images. Astrophys. J. **283**, 373–384 (1984)

Howard, R., Harvey, J.: Spectroscopic determinations of solar rotation. Solar Phys. **12**(1), 23–51 (1970)

Howard, R.: Solar rotation. Annu. Rev. Astron. Astrophys. **22**, 131–155 (1984)

Howard, R.: The rotation of the sun. Sci. Am. **232**, 106–114 (1975)

Howe, R.: Solar interior rotation and its variation. Living Rev. Solar Phys. **6**(1), 1 (2009)

Husak, M., Brajša, R., Špoljarić, D., Krajnović, D., Ruždjak, D., Skokić, I., Roša, D., Hržina, D.: Bošković's method for determining the axis and rate of solar rotation by observing sunspots in 1777. Solar Phys. **298**(10), 122 (2023)

Ikhsanov, R.N., Ivanov, V.G.: Properties of space and time distribution of solar coronal holes. Solar Phys. **188**(2), 245–258 (1999)

Insley, J.E., Moore, V., Harrison, R.A.: The differential rotation of the corona as indicated by coronal holes. Solar Phys. **160**(1), 1–18 (1995)

Javaraiah, J.: Long-term variations in the solar differential rotation. Solar Phys. **212**(1), 23–49 (2003)

Jejčič, S., Čadež, A.: The comparison of spectroscopic measurements of the solar rotation. Hvar Obs. Bull. **29**, 11–20 (2005)

Karachik, N., Pevtsov, A.A., Sattarov, I.: Rotation of solar corona from tracking of coronal bright points. Astrophys. J. **642**(1), 562–567 (2006)

Kariyappa, R.: Solar coronal rotation determined by X-ray bright points in Hinode/XRT and Yohkoh/SXT full-disc images. Astron. Astrophys. **488**(1), 297–301 (2008)

Khutsishvili, E.V., Gigolashvili, M.Sh., Kvernadze, T.M.: Differential rotation of the sun determined tracing sunspots and oscillations of sunspot tilt angle. Solar Phys. **206**(2), 219–228 (2002)

Komm, R., Corbard, T., Durney, B.R., González Hernández, I., Hill, F., Howe, R., Toner, C.: Solar subsurface fluid dynamics descriptors derived from global oscillation network group and Michelson Doppler imager data. Astrophys. J. **605**, 554–567 (2004)

Komm, R.W., Howard, R.F., Harvey, J.W.: Rotation rates of small magnetic features from two-dimensional and one-dimensional cross-correlation analyses. Solar Phys. **145**(1), 1–10 (1993)

Kosovichev, A.G., Schou, J., Scherrer, P.H., Bogart, R.S., Bush, R.I., Hoeksema, J.T., Aloise, J., Bacon, L., Burnette, A., de Forest, C., Giles, P.M., Leibrand, K., Nigam, R., Rubin, M., Scott, K., Williams, S.D., Basu, S., Christensen-Dalsgaard, J., Dappen, W., Jr. Rhodes, E.J., Jr. Duvall, T.L., Howe, R., Thompson, M.J., Gough, D.O., Sekii, T., Toomre, J., Tarbell, T.D., Title, A.M., Mathur, D., Morrison, M., Saba, J.L.R., Wolfson, C.J., Zayer, I., Milford, P.N.: Structure and rotation of the solar interior: initial results from the MDI medium-L program. Solar Phys. **170**(1), 43–61 (1997)

Kozuka, Y., Watanabe, Ta., Kojima, M., Ohyama, M., Saito, T.: Rotation rates of soft x-ray coronal structures. In: Proceedings of Kofu Symposium, pp. 359–362, July 1994

Kueveler, G., Woehl, H.: Daily variations of the photospheric equatorial rotation velocity of the Sun and its absolute values in 1981 and 1982 as determined from measurements using a two-dimensional photodiode array. Astron. Astrophys. **123**(1), 29–32 (1983)

Li, K.J., Xu, J.C.: The differential rotation of the chromosphere and the quiet chromosphere in the falling and rising periods of a solar cycle. Mon. Not. **528**(2), 1438–1444 (2024)

Livingston, W., Jr. Duvall, T.L.: Solar rotation, 1966 1978. Solar Phys. **61**(2), 219–231 (1979)

Livingston, W.C.: On the differential rotation with height in the solar atmosphere. Solar Phys. **9**(2), 448–451 (1969)

Lustig, G., Woehl, H.: Solar equatorial plasma rotation from 1983 until 1986. Astron. Astrophys. **218**(1–2), 299–301 (1989)

Lustig, G.: Solar rotation 1947–1981 - Determined from sunspot data. Astron. Astrophys. **125**(2), 355–358 (1983)

Lustig, G.: The sun's rotation derived from sunspots 1970–1979. Astron. Astrophys. **106**(1), 151 (1982)

Nash, A.G., Jr. Sheeley, N.R., Wang, Y.M.: Mechanisms for the rigid rotation of coronal holes. Solar Phys. **117**(2), 359–389 (1988)

Nash, A.G.: Coronal rotation at 3.5 R_{solar} determined from solwind data. Adv. Space Res. **11**(1), 47–50 (1991a)

Nash, A.G.: Rotation of the sun's outer white-light corona determined from SOLWIND data. Astrophys. J. **366**, 592 (1991b)

Navarro-Peralta, P., Sanchez-Ibarra, A.: An observational study of coronal hole rotation over the sunspot cycle. Solar Phys. **153**(1–2), 169–178 (1994)

Obridko, V.N., Shelting, B.D.: Coronal holes as indicators of large-scale magnetic fields in the corona. Solar Phys. **124**(1), 73–80 (1989)

Perez Garde, M., Vazquez, M., Schwan, H., Woehl, H.: Large-scale solar motions as determined by Doppler shift measurements using a linear photodiode array. Astron. Astrophys. **93**(1-2), 67–70 (1981)

Poljančić Beljan, I., Jurdana-Šepić, R., Brajša, R., Sudar, D., Ruždjak, D., Hržina, D., Pötzi, W., Hanslmeier, A., Veronig, A., Skokić, I., Wöhl, H.: Solar differential rotation in the period 1964-2016 determined by the Kanzelhöhe data set. Astron. Astrophys. **606**, A72 (2017)

Pulkkinen, P., Tuominen, I.: Velocity structures from sunspot statistics in cycles 10 to 22. II. Latitudinal velocity and correlation functions. Astron. Astrophys. **332**, 755–760 (1998)

Ruždjak, D., Brajša, R., Sudar, D., Skokić, I., Poljančić Beljan, I.: A relationship between the solar rotation and activity analysed by tracing sunspot groups. Solar Phys. **292**(12), 179 (2017)

Rybak, J.: On the rotation period of the green corona. Hvar Obs. Bull. **24**(1), 135 (2000)

Rybak, J.: Rotational characteristics of the green solar corona: 1964 1989. Solar Phys. **152**(1), 161–166 (1994)

Schou, J., Antia, H.M., Basu, S., Bogart, R.S., Bush, R.I., Chitre, S.M., Christensen-Dalsgaard, J., Di Mauro, M.P., Dziembowski, W.A., Eff-Darwich, A., Gough, D.O., Haber, D.A., Hoeksema, J.T., Howe, R., Korzennik, S.G., Kosovichev, A.G., Larsen, R.M., Pijpers, F.P., Scherrer, P.H., Sekii, T., Tarbell, T.D., Title, A.M., Thompson, M.J., Toomre, J.: Helioseismic studies of differential rotation in the solar envelope by the solar oscillations investigation using the Michelson Doppler imager. Astrophys. J. **505**, 390–417 (1998)

Schroeter, E.H., Woehl, H.: Differential rotation and giant cell circulation of solar Ca^+-network. Solar Phys. **49**(1), 19–32 (1976)

Schroeter, E.H.: The solar differential rotation - present status of observations. Solar Phys. **100**, 141 (1985)

Singh, V.K., Chandra, S., Thomas, S., Sharma, S.K., Vats, H.O.: Radial differential rotation of solar corona using radio emissions. Mon. Not. **505**(1), L16–L20 (2021)

Skokić, I., Sudar, D., Brajša, R.: Measuring differential rotation from full disk ALMA images of the Sun. Central Eur. Astrophys. Bull. **46**(4), 1–12 (2022)

Snodgrass, H.B.: Separation of large-scale photospheric Doppler patterns. Solar Phys. **94**(1), 13–31 (1984)

Sudar, D., Brajša, R., Skokić, I., Poljančić Beljan, I., Wöhl, H.: Meridional motion and Reynolds stress from Debrecen photoheliographic data. Solar Phys. **292**(7), 86 (2017)

Sudar, D., Brajša, R., Skokić, I., Temmer, M., Saar, S.: Rotation profiles of coronal bright points inside and outside of coronal holes. Solar Phys. **299**, 50 (2024)

Sudar, D., Brajša, R.: Solar rotation in the period 1611–1631 determined using observations of Christoph Scheiner. Solar Phys. **297**(10), 132 (2022)

Sudar, D., Saar, S.H., Skokić, I., Poljančić Beljan, I., Brajša, R.: Meridional motions and Reynolds stress from SDO/AIA coronal bright points data. Astron. Astrophys. **587**, A29 (2016)

Sudar, D., Skokić, I., Brajša, R., Saar, S.H.: Steps towards a high precision solar rotation profile: results from SDO/AIA coronal bright point data. Astron. Astrophys. **575**, A63 (2015)

Sudar, D., Skokić, I., Ruždjak, D., Brajša, R., Wöhl, H.: Tracing sunspot groups to determine angular momentum transfer on the Sun. Mon. Not. **439**(3), 2377–2385 (2014)

Wan, M., Gao, P.X., Xu, J.C., Shi, X.J., Xiang, N.B., Xie, J.L.: Differential rotation: the chromosphere to the quiet chromosphere. Mon. Not. **520**(1), 988–993 (2023)

Wan, M., Li, K.: Differential rotation of solar filaments early observed by the OGAUC in the period 1929–1941. Solar Phys. **297**(9), 126 (2022)

Wang, Y.M., Jr. Sheeley, N.R., Nash, A.G., Shampine, L.R.: The quasi-rigid rotation of coronal magnetic fields. Astrophys. J. **327**, 427 (1988)

Wang, Y.M., Jr. Sheeley, N.R.: Understanding the rotation of coronal holes. Astrophys. J. **414**, 916 (1993)

Weber, M.A., Acton, L.W., Alexander, D., Kubo, S., Hara, H.: A method for characterizing rotation rates in the soft x-ray corona. Solar Phys. **189**(2), 271–288 (1999)

Wöhl, H., Brajša, R., Hanslmeier, A., Gissot, S.F.: A precise measurement of the solar differential rotation by tracing small bright coronal structures in SOHO-EIT images. Results and comparisons for the period 1998-2006. Astron. Astrophys. **520**, A29 (2010)

Zaatri, A., Wöhl, H., Roth, M., Corbard, T., Brajša, R.: Comparison of the sidereal angular velocity of subphotospheric layers and small bright coronal structures during the declining phase of solar cycle 23. Astron. Astrophys. **504**(2), 589–594 (2009)

Chapter 7
Can We Explain and Understand the Solar Differential Rotation?

In this chapter, we try to answer the question of how the solar differential rotation can be explained and understood. The first question is what can be the causes of the solar differential rotation and the angular momentum transport from poles towards the equator. Two candidates are discussed in more detail: meridional motions and Reynolds stresses. Then the observational evidence of the angular momentum transport is given. Taking into account several experimental results, it will be shown that meridional motions taken alone cannot be the cause of the solar differential rotation, but Reynolds stresses represent indeed a plausible explanation of the solar differential rotation phenomenon.

7.1 What Can Be the Causes of the Solar Differential Rotation?

In previous chapters, we have seen how the solar rotation changes with the latitude and depth. In the next chapter we will discuss temporal variation of the solar rotation. So, the solar rotation is a function of the latitude, the depth, and the time. On the surface of the Sun and in other atmospheric layers a gradual increase of the angular rotation velocity from the poles towards the equator is observed. However, the polar region of the Sun is a special case, where other behavior was observed in some measurements, indicating a rigid rotation component at high latitudes. As a first approximation we consider the solar differential rotation as a symmetric phenomenon with respect to the solar equator. We will later see that a small, but significant deviation from this symmetry was observed, revealing a so-called north–south asymmetry of the solar rotation.

We start with the observational result that angular rotation velocity is increasing from the polar and high latitude regions towards the equator and that this phenomenon

R. Brajša and A. Hanslmeier, *Solar Rotation*, UNITEXT for Physics,
https://doi.org/10.1007/978-981-97-6879-0_7

is symmetric on both sides of the solar equator. Generally speaking, this experimental result can be comprehended on three different levels, in increasing order of complexity:

- to explain,
- to understand, and
- to interpret.

Described in a very simplified way, to explain means to relate the cause and effect, to understand means to relate the explanation with known laws of physics, and to interpret means to put the understanding in a broader context and to ask the question of why it is so. This line of reasoning can be applied for comprehension of any physical phenomenon, as described in Benz (2016). We note that here we use a slightly different terminology.

The main goal of this chapter is a modest one; we try to explain and, partly, to understand the complex phenomenon of solar differential rotation. We start with a very brief theoretical summary, introducing the context and without going into details, and then we search for observational evidence for a couple of possible mechanisms.

The observed higher angular rotation velocity in the equatorial region in comparison with middle and high solar latitudes requires a mechanism for transport of angular momentum from poles towards the equator. Angular momentum **L** is a physical quantity which is the vector product of the arm **r** (distance from the axis of rotation) and the linear momentum **p** (which is a product of the mass m and velocity **v**):

$$\boxed{\mathbf{L} = \mathbf{r} \times \mathbf{p}\ .} \tag{7.1}$$

The main condition which any mechanism for angular momentum transport should obey is the physical law of conservation of angular momentum.

The complex phenomenon of the solar differential rotation can be comprehended as a result of convective flows driven radially by the buoyancy force and deflected horizontally by the Coriolis force due to solar rotation. The well-known Coriolis force can be defined with the Coriolis acceleration:

$$\boxed{\mathbf{a}_\mathrm{c} = -2\omega \times \mathbf{v}\ ,} \tag{7.2}$$

where ω is the velocity of rotation and **v** the linear velocity in the frame of reference of the rotating body.

Now we list possible physical phenomena and mechanisms which can cause the solar differential rotation:

- meridional motions;
- correlation of velocity components, Reynolds stresses;
- anisotropic viscosity;
- Rossby waves;
- giant convection cells;

- solar wind torques;
- pole–equator temperature difference;
- latitude-dependent heat transport.

It is important to note that this list is not exhaustive and that often a combination of mechanisms is needed to explain and reproduce the observed profile of the solar differential rotation. We briefly describe some of these points, while some will be discussed in more detail later and the rest is self-evident.

The anisotropic viscosity means that there are different transport rates in different directions, caused by the presence of gravity. The Rossby waves are a specific type of inertial waves occurring naturally in rotating fluids. They govern the large-scale dynamics of a rotating sphere, which is extensively studied in the Earth's atmosphere and oceans (see, e.g., Ruždjak et al. (2023) and references therein). The giant convection cells, or simply giant cells, are an extrapolation of the granulation and supergranulation phenomena to the larger scales. While the existence of the granulation and supergranulation cells is well established, giant cells are still elusive, although there are quite a few observational indications that they might be real. The solar wind torques might remove the angular momentum from the polar regions of the Sun, which decelerates solar regions at high latitudes. The consequence would be effectively higher angular rotation velocity at low latitudes in accordance with the observed equatorial acceleration.

Further, models of the solar differential rotation can be divided into the axisymmetric models and non-axisymmetric models. Finally, development of the computer techniques in the last couple of decades enabled a rapid development of the so-called explicit numerical models of solar differential rotation. We will not describe such models in this book, unless their final results could be directly compared with observations presented here. The reader is referred to theoretical reviews about solar differential rotation for further information about this subtopic.

Theory of the solar differential rotation is a prominent challenge for astrophysicists and there exists extensive literature about this topic. Here we mention only a small sample of reviews and books for further reading: Howard (1975), Gilman (1981, 1992), Stix (1989, 2002), Ossendrijver (2003), Rüdiger and Hollerbach (2004), Miesch (2005), and Foukal (2013).

For some of the listed phenomena, the observational evidence remains elusive and some others should be excluded because a detailed calculation does not provide agreement with observations. So, we continue only with the first two mechanisms, namely meridian motions and Reynolds stresses.

7.1.1 Transport of Angular Momentum in a Nutshell

An excellent summary review of the angular momentum transport in the context of the large-scale flows in the solar convection zone can be found in Brun and Rempel (2009). Here we follow that presentation.

First we make the mean field decomposition of the velocity field: $\mathbf{v} = \langle \mathbf{v} \rangle + \mathbf{v}'$, where $\langle \mathbf{v} \rangle$ represents the mean flow and $\mathbf{v}'$ the fluctuating part. In a spherical coordinate system it is most convenient to decompose the equations into an angular momentum transport equation, the ϕ component of momentum transport, and the longitudinal component of the vorticity equations that combines both meridional components of the momentum equations. Using this decomposition we have the system of equations with $\langle \mathbf{v}_\phi \rangle = \omega r \sin\theta$:

$$\rho\,(r\sin\theta)^2\,\frac{\partial\omega}{\partial t} = -\nabla\cdot\mathbf{F} \tag{7.3}$$

$$\frac{\partial\langle\omega_\phi\rangle}{\partial t} + [...] = r\sin\theta\,\frac{\partial\omega^2}{\partial z} - \frac{g}{c_p r}\,\frac{\partial s}{\partial\theta}. \tag{7.4}$$

The symbols in the above equations have the following meanings:

- ρ—the density;
- r—the radius;
- θ—the colatitude;
- ω—the angular rotation velocity;
- t—the time;
- **F**—the angular momentum flux;
- [...]—the advection terms, meridional components of the Reynolds stress, and magnetic terms, omitted here to keep the presentation simple and clear;
- z—is oriented along the direction of the axis of rotation;
- g—the gravitational acceleration;
- c_p—the heat capacity at constant pressure;
- s—the specific entropy.

The most important terms in Eq. (7.4) are the two terms on the right-hand side describing a thermal wind balance of the differential rotation. The first term arises from the Coriolis force and the second term from pressure and buoyancy forces. In the absence of any entropy variation in latitude, ω has to be approximately constant on cylinders (the Taylor–Proudman theorem). In the angular momentum equation (7.3) the angular momentum flux **F** has the following components:

$$\boxed{F_i = r\sin\theta\,\rho\left\{\underbrace{\langle v_i\rangle\,\omega\,r\sin\theta}_{\text{Meridional flow}} + \underbrace{\langle v_i'\,v_\phi'\rangle}_{\text{Reynolds stress}} - \frac{1}{4\pi\rho}\underbrace{(\langle B_i'\,B_\phi'\rangle + \langle B_i\rangle\langle B_\phi\rangle)}_{\text{Maxwell stress}}\right\}.} \tag{7.5}$$

So, the angular momentum can be transported by the following mechanisms:

- large-scale meridional motions: $\langle v_r\rangle$, $\langle v_\theta\rangle$;
- Reynolds stresses: correlations of small-scale turbulent motions;
- Maxwell stresses: contribution of the magnetic field.

Additionally, the angular momentum can be transported through the viscous effects, which are omitted here, since in the Sun they are small and can be neglected. We note that the magnetic field is also decomposed into the mean and fluctuating parts $\mathbf{B} = \langle \mathbf{B} \rangle + \mathbf{B}'$. So, the Maxwell stress is separated into the contributions from the mean field and fluctuating field.

In the rest of this chapter we will discuss further only the two main contributors, namely meridional motions and Reynolds stresses. We will first connect theoretical concepts with the quantities which can be measured and then search for observational evidence of those observables. The influence of the magnetic field through the Maxwell stress will be treated in the next chapter. Note that the sign of the Maxwell stress is opposite to the signs of the other two terms on the right-hand side of Eq. (7.5).

Meridional motions and Reynolds stresses are closely related, although they contribute to the angular momentum transport differently. For further reading, we recommend the following publications: Käpylä (2006), Schmidt (1981, 1982), Ruediger (1980, 1982, 1989), and Durney (1976, 1989, 1991, 1996, 2000, 2001), representing only a small selection from the extensive literature describing theoretical aspects of the solar differential rotation and the angular momentum transport.

7.1.2 Meridional Motions: A Brief Theoretical Background

The large-scale meridional flow can transport angular momentum either to the equator or to the pole of the Sun. We repeat Eq. (7.1) which defines the angular momentum:

$$\boxed{\mathbf{L} = \mathbf{r} \times \mathbf{p}\ ,} \tag{7.6}$$

and consider a global, systematic meridional flow in either of the two directions:

- pole $\Rightarrow$ equator;
- equator $\Rightarrow$ pole.

We can visualize the following flow pattern: a global meridional motion from the pole to the equator at the solar surface at r_1, with a circulation back to the pole deep in the convection zone at r_2. Now the important point is that if $r_1 > r_2$, the transport of angular momentum towards the equator could take place, according to Eq. (7.6). This is because more angular momentum is transported if the flow happens at larger radius r. However, in the real Sun the described meridional circulation pattern can transport angular momentum towards equator only if one of the following physical conditions is fulfilled (Gilman 1981):

- A sufficient diffusion should be present to link the different layers of fluid and keep the angular velocity nearly constant, most likely associated with small-scale motions in the convection zone.

- The eddy viscosity should be anisotropic, meaning that the transport rates are different in different directions, due to the presence of gravity.
- The assumption that the convective heat flux is weakly influenced by rotation, so that it becomes a function of the latitude.

One of the main aims of this and the next subsection is to establish a link between theoretical considerations and practical solar observations, providing a transfer of physical quantities from equations into observables. Meridional motions are large-scale, longitudinally averaged, north–south velocity components on the Sun which lie in the meridional planes. They consist of the radial (v_r) and the latitudinal component (v_θ):

$$\boxed{\mathbf{v} = v_r(r,\theta)\,\mathbf{e}_r + v_\theta(r,\theta)\,\mathbf{e}_\theta\ ,} \tag{7.7}$$

where $\mathbf{e}_r$ and $\mathbf{e}_\theta$ are the unit vectors in the spherical polar coordinate system (r, θ, ϕ), where r is the radius, θ is the colatitude, and ϕ is the longitude. To avoid possible misunderstanding, it should be stressed that the latitudinal component of meridional motion, v_θ, represents the motion along the meridians and across the latitudes. Using tracers on the solar surface, only the latitudinal component, v_θ, can be measured (see, e.g., Brajsa and Woehl (2000)).

7.1.3 Reynold Stresses: A Brief Theoretical Background

Reynolds stresses are non-axisymmetric motions defined as correlations between the east–west and either the north–south or radial motions (Gilman 1981). Although the motions themselves are non-axisymmetric, in practical measurement the average is performed along the longitude to get a net correlation which leads to the angular momentum flux transport towards equator, even if there is no net mass flux. The needed tilts of the velocity vectors are caused by the Coriolis force acting on the convective motions.

In the hydrodynamical analysis of the solar differential rotation, an important question is whether the turbulence is caused by the shear, or vice versa, the shear, manifesting as the differential rotation, is caused by the turbulence. In Canuto et al. (1994) it is argued that the latter is the case on the Sun, and assuming the following cause and effect chain:

- buoyancy $\Rightarrow$ turbulence $\Rightarrow$ mean flow with differential rotation,

the so-called algebraic Reynolds stress models were developed. For our study, it is noteworthy that such models predict a latitude-dependent horizontal component of the Reynolds stress, which can be compared with measurements. Further theoretical works which present models of latitude-dependent Reynolds stresses include Hupfer (2005) and Hupfer et al. (2005).

The components of the Reynolds stress tensor from Eq. (7.5) in the spherical polar coordinate system (r, θ, ϕ) are

$$\boxed{Q_{r\phi} = \langle v_r v_\phi \rangle} \tag{7.8}$$

$$\boxed{Q_{\theta\phi} = \langle v_\theta v_\phi \rangle}. \tag{7.9}$$

Here $\langle v_i v_j \rangle$ represents averaging over longitudes and time, for a given latitude. Obviously, a component of the Reynolds stress is measured in $m^2 s^{-2}$, since the velocity components are given in m/s. It is important to note that using tracers at the solar surface only the so-called horizontal component of the Reynolds stress $Q_{\theta\phi}$ can be measured.

Finally, we should transfer the horizontal component of the Reynolds stress $Q_{\theta\phi}$ into the observables which can be measured. The velocity components in the spherical polar coordinate system should be identified in the natural coordinate system on the solar surface, widely used by the observers. First, the angular rotation velocities (v_θ and v_ϕ, expressed in, e.g., deg/day) should be converted into the linear rotation velocities (v_{mer} and v_{rot}, expressed in m/s). Meridional velocity and rotational velocity are denoted by v_{mer} and v_{rot}, respectively. Then follows the mean field decomposition of the velocity field: $v_i = \langle v_i \rangle + v_i{}'$, where $\langle v_i \rangle$ represents the mean flow and $v_i{}'$ the fluctuating part.

In our case the horizontal component of the Reynolds stress $Q_{\theta\phi}$ is

$$\boxed{Q = \langle v_{\mathrm{mer}} v_{\mathrm{rot}} \rangle\ ,} \tag{7.10}$$

and with the following decomposition:

$$v_{\mathrm{mer}} = \langle v_{\mathrm{mer}} \rangle + v'_{\mathrm{mer}} \tag{7.11}$$

$$v_{\mathrm{rot}} = \langle v_{\mathrm{rot}} \rangle + v'_{\mathrm{rot}}\ , \tag{7.12}$$

we get

$$Q = \langle v_{\mathrm{mer}} \cdot v_{\mathrm{rot}} \rangle = \langle \langle v_{\mathrm{mer}} \rangle + v'_{\mathrm{mer}} \rangle \cdot \langle \langle v_{\mathrm{rot}} \rangle + v'_{\mathrm{rot}} \rangle \tag{7.13}$$

$$Q = \langle \langle v_{\mathrm{mer}} \rangle \cdot \langle v_{\mathrm{rot}} \rangle \rangle + \langle \langle v_{\mathrm{mer}} \rangle \cdot v'_{\mathrm{rot}} \rangle + \langle v'_{\mathrm{mer}} \cdot \langle v_{\mathrm{rot}} \rangle \rangle + \langle v'_{\mathrm{mer}} \cdot v'_{\mathrm{rot}} \rangle \tag{7.14}$$

$$Q = \langle v_{\text{mer}} \rangle \cdot \langle v_{\text{rot}} \rangle + \langle v_{\text{mer}} \rangle \cdot \underbrace{\langle v'_{\text{rot}} \rangle}_{=0} + \underbrace{\langle v'_{\text{mer}} \rangle}_{=0} \cdot \langle v_{\text{rot}} \rangle + \langle v'_{\text{mer}} \cdot v'_{\text{rot}} \rangle \tag{7.15}$$

$$Q = \langle v_{\text{mer}} \rangle \cdot \langle v_{\text{rot}} \rangle + \langle v'_{\text{mer}} \cdot v'_{\text{rot}} \rangle \ . \tag{7.16}$$

In the above derivation, the so-called Reynolds rules were used. For the angular momentum transport the second term in Eq. (7.16) is important. Now we shall relate these fluctuating velocity components with observable quantities on the Sun:

$$\mid \langle v_{\text{rot}} \rangle \mid >> \sqrt{\mid \langle {v'_{\text{rot}}}^2 \rangle \mid} \Rightarrow v'_{\text{rot}} = \Delta v_{\text{rot}} \tag{7.17}$$

$$\mid \langle v_{\text{mer}} \rangle \mid << \sqrt{\mid \langle {v'_{\text{mer}}}^2 \rangle \mid} \Rightarrow v'_{\text{mer}} = v_{\text{mer}} \ , \tag{7.18}$$

where Δv_{rot} is the rotation velocity residual, i.e., the deviation of individual rotation velocity from the mean value and v_{mer} is the meridional velocity. In this way we finally get the horizontal Reynolds stress expressed in terms of observables of the solar rotation measurements:

$$\boxed{Q = \langle \Delta v_{\text{rot}} v_{\text{mer}} \rangle \ .} \tag{7.19}$$

7.2 Observational Evidence of the Angular Momentum Transport

In the rest of this chapter, we present some observational results on meridional motions and Reynolds stresses based on tracing sunspots and coronal bright points. The main question is whether these results support the concept of angular momentum transport for reproducing the observed differential rotation profile of the Sun.

7.2.1 Meridional Motions: Results of Measurements

Meridional velocity is usually measured as the tracer's difference in latitude b over elapsed time t:

$$\boxed{v_{\text{mer}} = \frac{\Delta b}{\Delta t} \ .} \tag{7.20}$$

The next step is the conversion of the velocity values from deg/day to m/s, knowing the solar radius. Finally, it should be decided to which latitude the calculated meridional velocity vector should be ascribed, since we have two measurements of position for one velocity value. The meridional velocity vector should be assigned to the latitude of the first position measurement (Sudar et al. 2014 and references therein). Then there is no net flow into the latitude bin from other latitudes and we do not have to worry about the non-uniform distribution of sunspots in latitudes.

There is extensive literature about the measurements of meridional motions on the Sun. We note that in many cases the results are contradictory, partly due to systematic and methodological errors in data reduction. Results published previous to about 2010 were summarized and discussed in Wöhl and Brajša (2001), Vršnak et al. (2003), Sudar et al. (2014, 2016, 2017), and Ruždjak et al. (2018). We continue this subsection by describing some recent results based on tracing sunspots (mainly sunspot groups) and coronal bright points.

Meridional motions of sunspot groups from the extended Greenwich data set (Sudar et al. 2014), from the Debrecen data set (Sudar et al. 2017), and from the Kanzelhöhe data set (Ruždjak et al. 2018) are presented in Figs. 7.1, 7.2, and 7.3, respectively. The results clearly indicate the existence of poleward motions in equatorial and low latitude regions. This direction reverses the sign at about 15–20 deg in latitude and in the middle latitudes, up to about 40 deg, the meridional flow towards equator is measured. At latitudes higher than 40 deg the meridional flow is again poleward, for two out of the three analyzed data sets. This result was obtained when data for both solar hemispheres are treated together and it is confirmed for the two

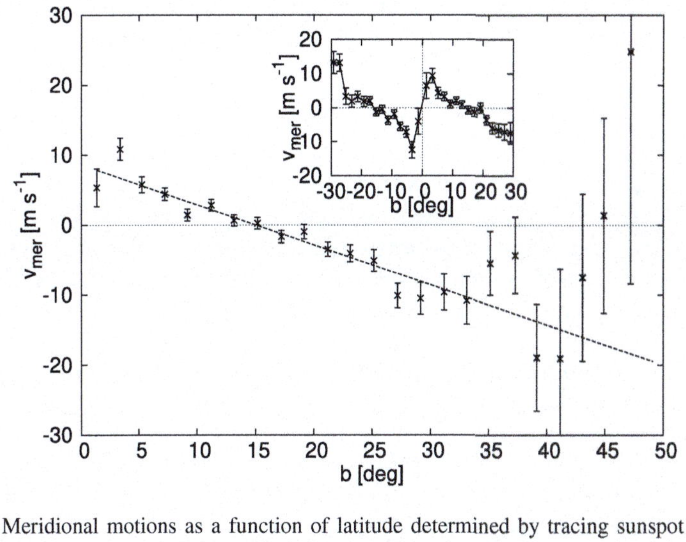

Fig. 7.1 Meridional motions as a function of latitude determined by tracing sunspot groups in the period 1878–2011 using the extended Greenwich data set. Positive velocity values refer to the motions towards the pole and negative velocity values indicate motions towards the equator. Average values for the 2 deg latitude bins are given with the error bars. The linear fit is denoted with the dashed line. Both solar hemispheres are treated together and in the inset the results for the two hemispheres are presented separately, with the sign reversed for the southern hemisphere. Credit: Sudar et al. (2014)

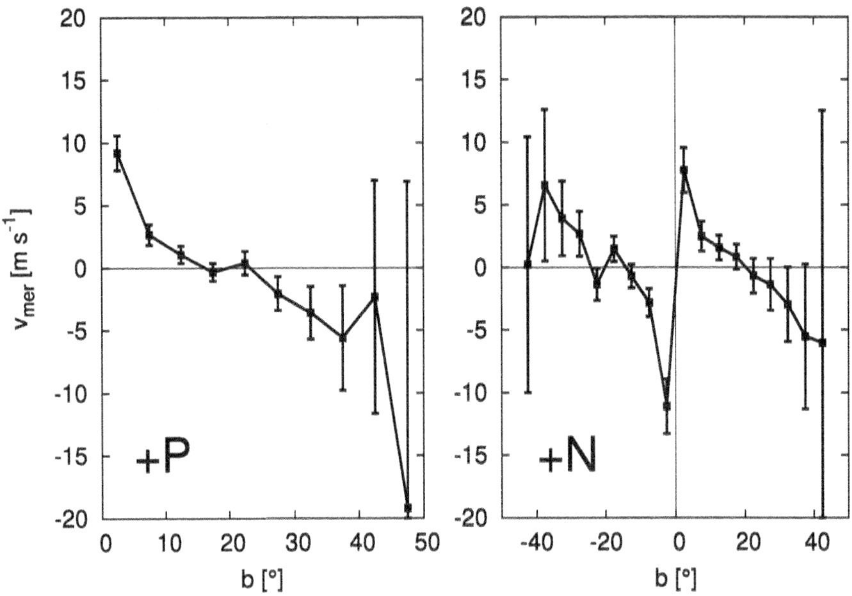

Fig. 7.2 Meridional motions as a function of the latitude for sunspot groups data from the Debrecen Photoheliographic Data catalogue for the period 1974–2016. Average values for the 5 deg latitude bins are given together with corresponding error bars. In the plot on the left-hand side both solar hemispheres are folded together and the positive velocity values indicate motions towards the poles (+P). In the plot on the right-hand side the two solar hemispheres are presented separately and the positive velocity values indicate motions towards the north pole of the Sun (+N). Credit: Sudar et al. (2017)

solar hemispheres taken separately. So, the pattern of meridional motions of sunspot groups is generally symmetric respectively to the solar equator.

It should be noted that the results on meridional motions presented in Figs. 7.1, 7.2, and 7.3 reflect the flow velocities averaged over several solar cycles. Moreover, the well-known butterfly diagram (systematic migration of sunspots from medium to lower latitudes on both hemispheres during a solar cycle) should be taken into account. We can now summarize the general picture of meridional motions determined by tracing sunspot groups and infer what are the implications of these observational results for theoretical models:

- There is no overall meridional flow determined by sunspots from poles to equator $\Rightarrow$ the observed meridional motions, considered alone, cannot explain the differential rotation of the Sun.
- Meridional motions on both solar hemispheres are directed towards the center of sunspot activity $\Rightarrow$ this observed flow pattern is consistent with the existence of large subphotospheric convective rolls, as hypothesized in Howard (1987), especially taking into account that sunspots bring rotational information from the levels below the photosphere.

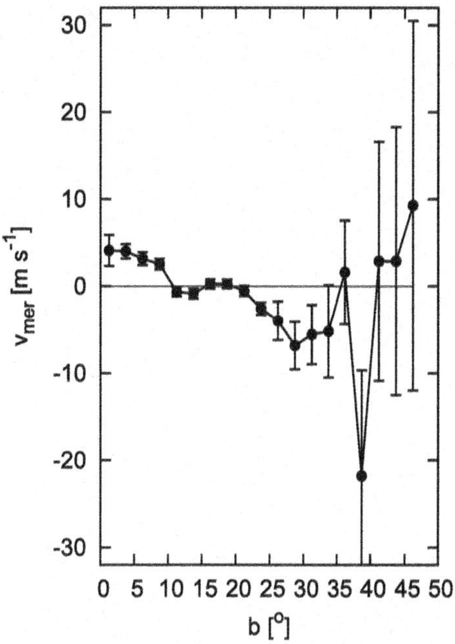

Fig. 7.3 Meridional motions as a function of the latitude for sunspot groups traced in the Kanzelhöhe drawings in the period 1964–2016. Average values are given for the 2.5 deg latitude bins together with the corresponding error bars. Both solar hemispheres are folded together and the positive velocity values indicate motions towards the poles and the negative ones towards the equator. Credit: Ruždjak et al. (2018)

Next we discuss the meridional flow pattern determined by tracing coronal bright points. The results of the five months data analysis of full-disk 19.3 nm images taken with SDO/AIA telescope are presented in Figs. 7.4 and 7.5. Both plots refer to the same data; in the first plot the whole data set is reduced and in the second plot the results for coronal bright points inside and outside of coronal holes are treated separately. The interpretation of the meridional flow pattern determined by tracing coronal bright points is even more straightforward than for the case of the sunspots. Here is the summary and the implications:

- There is no overall meridional flow determined by coronal bright points from poles to equator $\Rightarrow$ the observed meridional motions, considered alone, cannot explain the differential rotation of the Sun.
- There exists an overall meridional flow pattern from equator to the poles $\Rightarrow$ this observed flow pattern is consistent with the poleward meridional motions needed as input in the flux-transport solar dynamo model (e.g., Sanchez et al. (2014) and references therein) and in the model of the rigid rotation of coronal holes (Nash et al. 1988, Wang and Sheeley 1993). Moreover, in the latter case the amplitude of the observed meridional velocity is in a general agreement with the values assumed in the model.

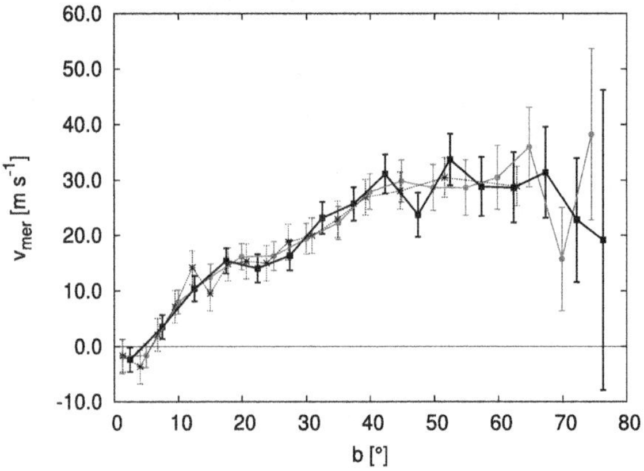

Fig. 7.4 Meridional motions as a function of the latitude obtained by tracing coronal bright points in the SDO/AIA 19.3 nm images in the period January–May 2011. Both solar hemispheres are folded together and positive velocity values indicate motions towards the poles. Average velocity values with corresponding error bars are shown for three sampling methods: for 5 deg latitude bins (thick black solid line and filled squares), for 2.5 deg latitude bins (gray solid line and filled circles), and for bins with the same number of data points ($n = 5000$, dotted line and stars). Credit: Sudar et al. (2016)

Finally, we raise the question of the non-axisymmetric meridional motions. Up to now all meridional motions were considered as symmetric along the latitude circles, i.e., averaged for all longitudes. However, there are several observational examples of the opposite behavior. In Fig. 7.6 a 3D-plot for meridional motions of coronal bright points traced in the full-disk solar images from the SOHO/EIT instrument at a wavelength of 28.4 is presented. The longitude is on the horizontal axis, the latitude on the vertical axis, and the third coordinate represents color coded meridional velocity. A significant deviation from the symmetry around the rotational axis is clearly visible. It is noteworthy that such non-axisymmetric meridional motions are consistent with several other observational studies, such as Schroeter and Woehl (1976), Brandt et al. (1988), and Ambroz (1993).

7.2.2 *Correlation of Velocity Components: Results of Measurements*

As we have seen in the previous subsection, the meridional motions, taken alone, cannot be the cause of the solar differential rotation with the observed equatorial acceleration. So, in the next step we search for the correlation between the relevant velocity components, the meridional velocity ($v_{\rm mer}$) and the rotation velocity residual ($\Delta v_{\rm rot}$). The question is whether the tracers with a larger velocity than average move towards the pole or towards the equator, again considered on the average.

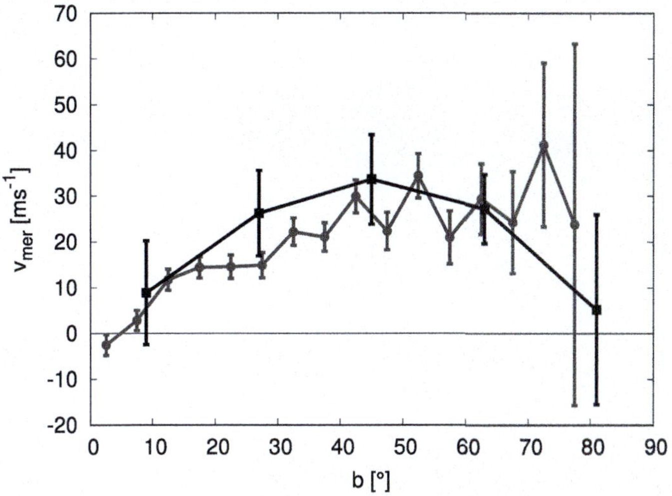

Fig. 7.5 Similar to Fig. 7.4, but presented separately for coronal bright points detected outside of coronal holes (gray) and for coronal bright points detected inside of coronal holes (black). The width of the respective latitude bins is different, due to the difference in the number of data points within the two data subsets (the number of coronal bright points inside coronal holes is smaller than the other sample). Positive velocities are directed towards the poles. Credit: Sudar et al. (2024)

The correlations between the two velocity components determined by tracing sunspot groups in the extended Greenwich data set, in the Debrecen Photoheliographic Data catalogue, and in the Kanzelhöhe drawings data set are presented in Figs. 7.7, 7.8, and 7.9, respectively. The data are fitted with a straight line and the least-square fitting procedure gives the following parameters with corresponding errors.

- The extended Greenwich data set, 1878–2011, Fig. 7.7, Sudar et al. (2014):

$$v_{\rm mer} = (-0.0804 \pm 0.0017)\, \Delta v_{\rm rot} + (-0.12 \pm 0.27)\, {\rm ms}^{-1} \,. \tag{7.21}$$

- The Debrecen Photoheliographic Data catalogue, 1974–2016, Fig. 7.8, Sudar et al. (2017):

$$v_{\rm mer} = (-0.0876 \pm 0.0021)\, \Delta v_{\rm rot} + (-1.01 \pm 0.34)\, {\rm ms}^{-1} \,. \tag{7.22}$$

- Kanzelhöhe drawings, 1964–2016, Fig. 7.9, Ruždjak et al. (2018):

$$v_{\rm mer} = (-0.0912 \pm 0.0028)\, \Delta v_{\rm rot} + (-0.42 \pm 0.43)\, {\rm ms}^{-1} \,. \tag{7.23}$$

The analyzed data sets embrace several solar cycles and partially overlap. The resulting equation for each data set is an average over the whole time period under investigation. From the figures, it can be seen that, indeed, tracers rotating faster (a

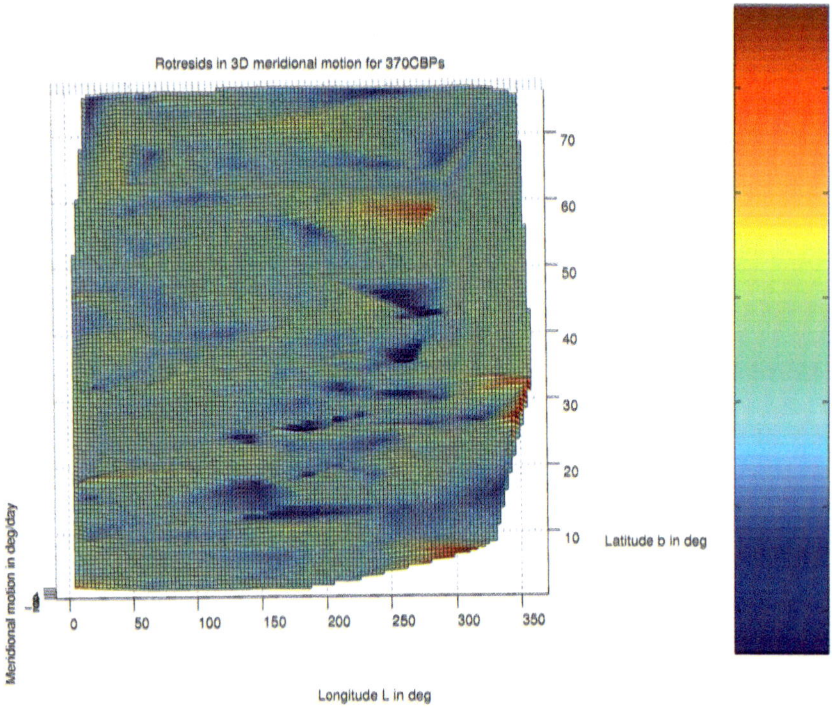

Fig. 7.6 A planar 3D-plot displaying a non-axisymmetric behavior of meridional motions on the Sun. The plot is based on tracing 370 coronal bright points in October/November 1999 on full-disk solar images taken with the SOHO/EIT instrument at the wavelength of 28.4 nm. On the x-axis is the solar longitude in degrees, on the y-axis is the latitude in degrees, and the third coordinate (color) represents the meridional velocity. The color bar denotes the meridional velocity in deg/day, displayed in colors spanning from blue to red. Positive meridional velocities represent motions to the poles. Credit: Mulec (2007)

positive rotation velocity residual) are moving on the average predominantly towards the equator (a negative meridional velocity). It is especially important to note the following characteristics of this correlation. It is a statistical phenomenon, similar to the whole solar differential rotation behavior. This means that one, single data point in these figures does not have very important physical meaning; what counts is the statistical behavior of all measured data resulting in the above equations. It is also noteworthy that slopes of the fitted straight lines have a rather high level of statistical significance, expressed as a ratio between the error and the coefficient defining the slope of the line.

We can conclude that the correlation between the rotation velocity residuals and meridional velocities shows that on the average:

- Objects which rotate faster than average are the same objects which move towards the equator.

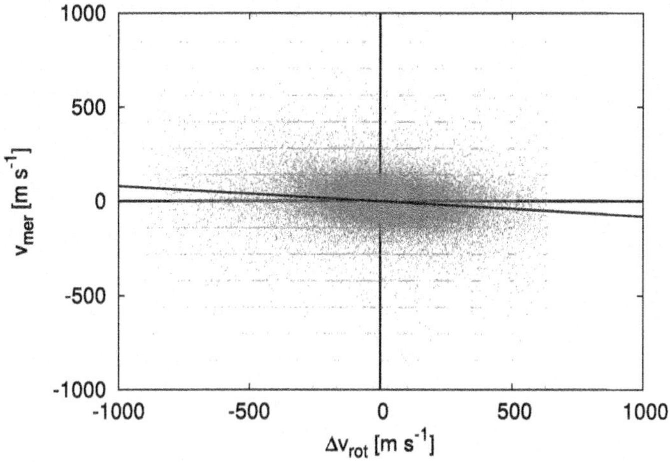

Fig. 7.7 Correlation of velocity components, meridional velocity (v_{mer}) and rotation velocity residual (Δv_{rot}), determined by tracing sunspot groups in the period 1878–2011 using the extended Greenwich data set. Positive meridional velocities refer to the motions towards the pole and negative meridional velocities refer to the motions towards the equator. Positive rotation velocity residuals denote faster rotation than average and negative rotation velocity residuals denote slower rotation than average. The linear fit is denoted with the full line and both solar hemispheres are treated together. Credit: Sudar et al. (2014)

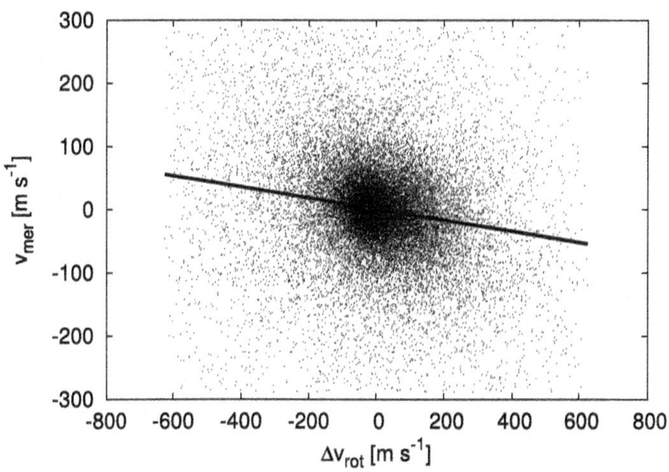

Fig. 7.8 Correlation of velocity components, meridional velocity (v_{mer}) and rotation velocity residual (Δv_{rot}), for sunspot groups data from the Debrecen Photoheliographic Data catalogue for the period 1974–2016. Positive meridional velocities refer to the motions towards the pole and negative meridional velocities refer to the motions towards the equator. Positive rotation velocity residuals denote faster rotation than average and negative rotation velocity residuals denote slower rotation than average. The linear fit is denoted with the full line and both solar hemispheres are treated together. Credit: Sudar et al. (2017)

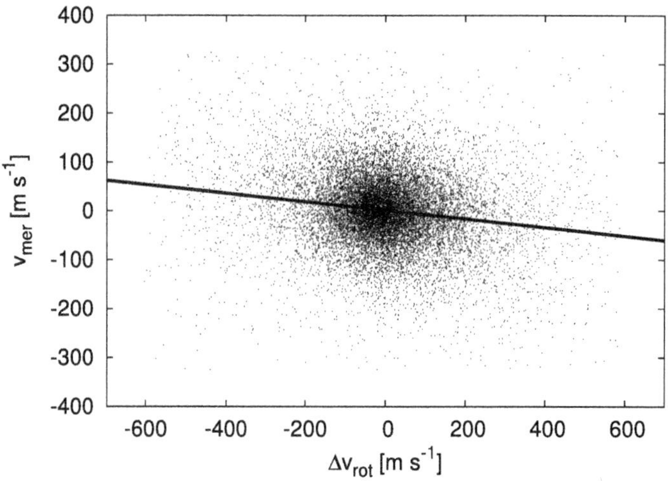

Fig. 7.9 Correlation of velocity components, meridional velocity (v_{mer}) and rotation velocity residual (Δv_{rot}), for sunspot groups traced in the Kanzelhöhe drawings in the period 1964–2016. Positive meridional velocities refer to the motions towards the pole and negative meridional velocities refer to the motions towards the equator. Positive rotation velocity residuals denote faster rotation than average and negative rotation velocity residuals denote slower rotation than average. The linear fit is denoted with the full line and both solar hemispheres are treated together. Credit: Ruždjak et al. (2018)

- Objects which rotate slower than average are the same objects which move towards the poles.

This is qualitatively consistent with the transport of angular momentum towards the equator, needed to explain the observed profile of the solar differential rotation. The final question is whether this angular momentum transport has the needed amount to maintain the observed difference in the angular velocity along the latitudes. We address this question in the next subsection.

7.2.3 Reynolds Stresses: Results of Measurements

To maintain the observed profile of the solar differential rotation theoretical models require the existence of the horizontal Reynolds stress of the order of several thousand $(\mathrm{m/s})^2$, see, e.g., Stix (1989, 2002) and Rüdiger and Hollerbach (2004). Of course, the transport of angular momentum should be towards the equator, which requires the sign of the horizontal Reynolds stress to be negative. Moreover, an increase in the absolute value of the horizontal Reynolds stress with increasing latitude, with inflection point at the middle and high latitudes, was predicted (e.g., Canuto et al. (1994)).

There are many indications that such Reynolds stress really exists on the Sun. We mention here quite a few research papers in which such observational results were reported before the year 2000, using various solar objects as tracers: Ward (1965), Belvedere et al. (1976), Schroeter and Woehl (1976), Gilman and Howard (1984), D'Silva and Howard (1995), Meunier et al. (1997), and Pulkkinen and Tuominen (1998). Moreover, in some of these studies a latitudinal dependence of the horizontal Reynolds stress component was also investigated.

We now summarize some of the representative results about this topic published more recently, after the year 2000. In Fig. 7.10 the horizontal Reynolds stress as a function of the latitude is presented for the coronal bright points traced in the 28.4 nm SOHO/EIT images in the period 1998–1999 and for tracing the sunspot groups in the extended Greenwich data set in the period 1878–2011. For each case two different representations of the data were used, regarding the sampling and fitting procedure, as written in figure caption. Both solar hemispheres are treated together and the negative sign of the horizontal Reynolds stress denotes the angular momentum transport towards solar equator. Earlier results obtained using a subset (1925–1954) of the Greenwich Photoheliographic Results are also given in this figure for comparison.

The dependence of the horizontal Reynolds stress Q on the latitude b has a high statistical significance and the linear least-square fit through all individual data points from the extended Greenwich data set (1878–2011) is according to Sudar et al. (2014) represented by

$$Q = (-76.4 \pm 9.5)(\mathrm{m/s})^{-2}(\mathrm{deg})^{-1}\, b + (-933 \pm 161)(\mathrm{m/s})^{-2}\ . \tag{7.24}$$

The straight line corresponding to this formula is given in Fig. 7.10 as described in the figure caption.

To investigate further dependence of the horizontal Reynolds stress on the latitude, two additional, high quality data sets of sunspot positions were used and the representation of the results was improved. The results for sunspot groups data from the Debrecen Photoheliographic Data catalogue for the period 1974–2016 and from the Kanzelhöhe drawings data set for the period 1964–2016 are given in Figs. 7.11 and 7.12, respectively. A qualitative agreement with previous results obtained by tracing sunspot groups from other data sets and coronal bright points was found. Further, in Figs. 7.11 and 7.12, the best-fit profile for the functional dependence of the horizontal Reynolds stress with the latitude is also given. Observational data are fitted according to the function (Sudar et al. 2017):

$$\boxed{Q = c_1\, b\, \mathrm{e}^{-c_3 b^2}\ ,} \tag{7.25}$$

where Q is the horizontal Reynolds stress, b is the latitude, and c_1 and c_3 are coefficients to be determined by fitting results of measurements. Equation (7.25) is valid only if the function is perfectly asymmetric around zero: $Q(b) = -Q(-b)$. Exam-

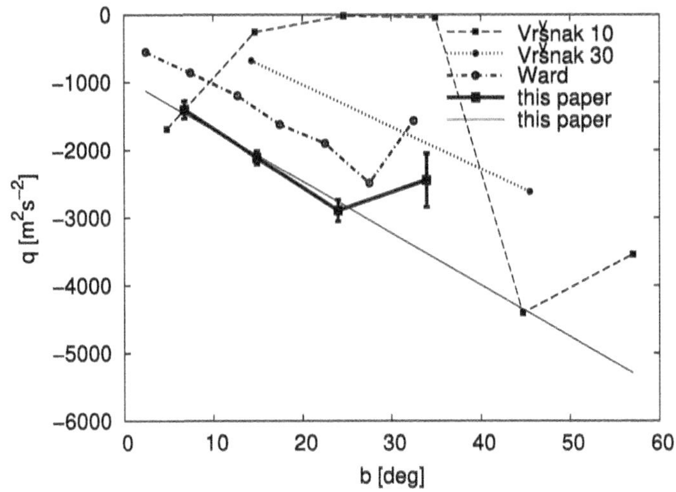

Fig. 7.10 Horizontal Reynolds stress $q = \langle \Delta v_{\text{rot}} v_{\text{mer}} \rangle$ as a function of the latitude b for various tracers, as listed below. Credit: Sudar et al. (2014). Notes: - Vršnak 10: coronal bright points traced in SOHO/EIT 28.4 nm images in 1998–1999, average values for 10 deg latitude bins, Vršnak et al. (2003) - Vršnak 30: coronal bright points traced in SOHO/EIT 28.4 nm images in 1998–1999, average values for 30 deg latitude bins, Vršnak et al. (2003) - Ward: sunspot groups, Greenwich Photoheliographic Results, 1925–1954, average values for 5 deg latitude bins, Ward (1965) - this paper: sunspot groups, extended Greenwich data set, 1878–2011, black squares: average values for 10 deg latitude bins, Sudar et al. (2014) - this paper: sunspot groups, extended Greenwich data set, 1878–2011, thin solid line: linear fit defined by Eq. (7.24) Sudar et al. (2014)

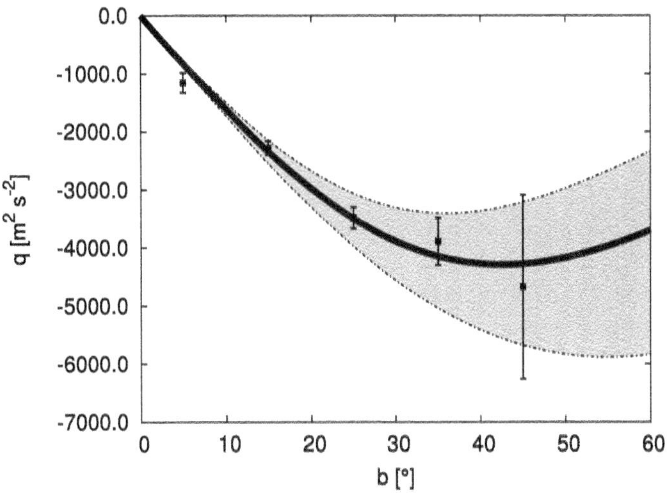

Fig. 7.11 Horizontal Reynolds stress $q = \langle \Delta v_{\text{rot}} v_{\text{mer}} \rangle$ as a function of the latitude b for sunspot groups data from the Debrecen Photoheliographic Data catalogue for the period 1974–2016. Black squares with error bars denote the average values for the 10 deg latitude bins. The best-fit profile for the function of the form given in Eq. (7.25) is also given. Shaded area refers to the range of possible values defined by the errors of the best-fit coefficients. Credit: Sudar et al. (2017)

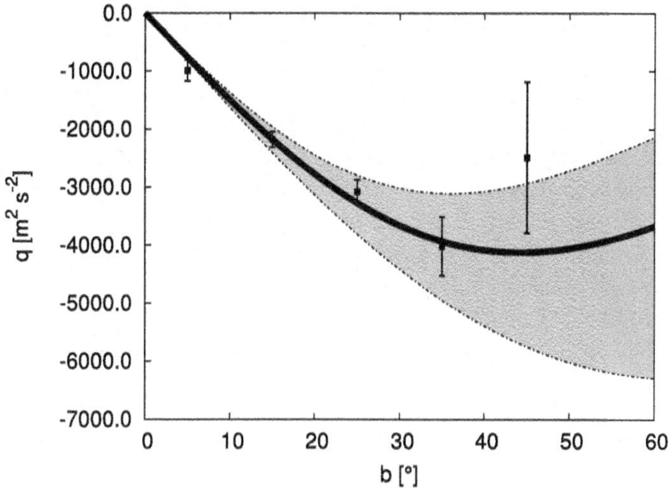

Fig. 7.12 Horizontal Reynolds stress $q = \langle \Delta v_{\rm rot} v_{\rm mer} \rangle$ as a function of the latitude b for sunspot groups traced in the Kanzelhöhe drawings in the period 1964–2016. Black squares with error bars denote the average values for the 10 deg latitude bins. The best-fit profile for the function of the form given in Eq. (7.25) is also given. Shaded area refers to the range of possible values defined by the errors of the best-fit coefficients. Credit: Ruždjak et al. (2018)

Table 7.1 The best-fit coefficients from Eq. (7.25) for the extended Greenwich data set (Sudar et al. 2017)

Coefficient	Value	Relative error
c_1 [m^2 s^{-2} deg^{-1}]	-166 ± 10	6.2%
c_3 [deg^{-2}]	0.00028 ± 0.00011	39.9%

Table 7.2 The best-fit coefficients from Eq. (7.25) for the Kanzelhöhe drawings data set (Ruždjak et al. 2018)

Coefficient	Value	Relative error
c_1 [m^2 s^{-2}deg^{-1}]	-154 ± 11	7.3%
c_3 [deg^{-2}]	0.00026 ± 0.00013	50.8%

ples of numerical fitting results for sunspot groups data from the Debrecen Photoheliographic Data catalogue for the period 1974–2016 and from the Kanzelhöhe drawings data set for the period 1964–2016 are given in Tables 7.1 and 7.2, respectively. It is noteworthy that fitting the observational results with a function of the type (7.25) enables identifying the inflection point in the curve $Q(b)$. Examples of the curves based on experimental data, as presented in Figs. 7.11 and 7.12, are in agreement with theoretical models which predict a gradual decrease of the Reynolds stress amplitude from medium to high solar latitudes (see, e.g., Canuto et al. (1994)).

7.2.4 Summary of Observational Results and Conclusion

Measurements of the latitudinal dependence of the covariance of rotation velocity residuals and meridional flow presented in this chapter strongly suggest that the Reynolds stress is the dominant mechanism which explains the observed characteristics of the solar differential rotation: equatorial acceleration and decrease of the angular velocity with increasing latitude on both solar hemispheres. The observed horizontal Reynolds stress is consistent with the transport of angular momentum towards the solar equator both qualitatively (correct sign) and quantitatively (absolute value and the dependence on the latitude). On the other hand, the model of axisymmetric meridional circulation, which could also produce transfer of the angular momentum towards the equator, as discussed earlier in this chapter, is strongly disfavored because it requires surface meridional flow in the direction towards the solar equator, which contradicts the results of measurements presented here. A comparison between observations and theory, including discrimination between various models, was discussed in more detail in Vršnak et al. (2003), Sudar et al. (2014, 2017), and Ruždjak et al. (2018).

Exercises and Questions

1. What are the levels of comprehension of astrophysical phenomena?

2. What is the role of angular momentum and angular momentum transport in efforts to explain the solar differential rotation?

3. Explain what is the Coriolis force and how it is acting in the northern and the southern hemispheres of a rotating body.

4. What are the possible physical mechanisms which can be the cause of the solar differential rotation? Explain how they might produce the observed rotational profile on the Sun.

5. What are the components of the angular momentum flux on the Sun?

6. Explain how the meridional motions and Reynolds stresses can transport angular momentum towards solar equator.

7. Find in the literature the Reynolds rules and prove the formula for the horizontal Reynolds stress using the observables from the solar rotation measurements.

8. What are the results of measurements of meridional motions and horizontal Reynolds stress for various tracers? Explain how the correlation of the measured velocity components leads to the concept of the horizontal Reynolds stress.

9. Discuss whether indications of non-axisymmetric meridional motions were found.

10. Describe the latitudinal dependence of the horizontal Reynolds stress and explain why it is important for comparison with theoretical models.

11. What is the state-of-the-art explanation of the solar differential rotation?

References

Ambroz, P.: Evolutionary characteristics of large-scale magnetic and velocity fields. In: H. Zirin, G. Ai, H. Wang (eds.) IAU Colloq. 141: The Magnetic and Velocity Fields of Solar Active Regions, Astronomical Society of the Pacific Conference Series, vol. 46, p. 495 (1993)

Belvedere, G., Godoli, G., Motta, S., Paterno, L., Zappala, R.A.: Surface Reynolds stresses determined from the analysis of facular motions and the maintenance of the sun's differential rotation. Solar Phys. **46**(1), 23–28 (1976)

Benz, A.: Astrophysics and Creation. The Crossroad Publ. Co. (2016)

Brajsa, R., Woehl, H.: On the meridional motions of recurrent stable sunspot groups. Hvar Obs. Bull. **24**(1), 125 (2000)

Brandt, P.N., Scharmer, G.B., Ferguson, S., Shine, R.A., Tarbell, T.D., Title, A.M.: Vortex flow in the solar photosphere. Nature **335**(6187), 238–240 (1988)

Brun, A.S., Rempel, M.: Large scale flows in the solar convection zone. Space Sci. Rev. **144**(1–4), 151–173 (2009)

Canuto, V.M., Minotti, F.O., Schilling, O.: Differential rotation and turbulent convection: a new Reynolds stress model and comparison with solar data. Astrophys. J. **425**, 303 (1994)

D'Silva, S., Howard, R.F.: Sunspot velocity correlations: are they due to Reynolds stresses or to the Coriolis force on rising flux tubes? Solar Phys. **159**(1), 63–88 (1995)

Durney, B.R.: Approximate isocontours for the solar angular velocity in the convection zone. Solar Phys. **202**(2), 201–228 (2001)

Durney, B.R.: Meridional motions and the angular momentum balance in the solar convection zone. Astrophys. J. **528**(1), 486–492 (2000)

Durney, B.R.: Observational constraints on theories of the solar differential rotation. Astrophys. J. **378**, 378 (1991)

Durney, B.R.: On the influence of gradients in the angular velocity on the solar meridional motions. Solar Phys. **169**(1), 1–32 (1996)

Durney, B.R.: On theories of solar rotation. In: Bumba, V., Kleczek, J. (eds.) Basic Mechanisms of Solar Activity, vol. 71, p. 243 (1976)

Durney, B.R.: Some controversial issues in theories of the solar differential rotation and dynamo. Solar Phys. **123**(2), 197–216 (1989)

Foukal, P.: Solar Astrophysics, 3rd. Revised Edition. Wiley-VCH-Varlag (2013)

Gilman, P.A., Howard, R.: On the correlation of longitudinal and latitudinal motions of sunspots. Solar Phys. **93**(1), 171–175 (1984)

Gilman, P.A.: Global circulation and the solar dynamo. In: Jordan, S. (ed.) NASA Special Publication, vol. 450, pp. 231–252 (1981)

Gilman, P.A.: What can we learn about solar cycle mechanisms from observed velocity fields? In: Harvey, K.L. (ed.) The Solar Cycle, Astronomical Society of the Pacific Conference Series, vol. 27, p. 241 (1992)

Howard, R.F.: Models of motions in the Sun. Nature **328**(6132), 667–668 (1987)

Howard, R.: The rotation of the Sun. Sci. Am. **232**, 106–114 (1975)

Hupfer, Ch.: Breitenabhängigkeit des Reynolds-Tensors, Dissertation, University of Freiburg, Germany (2005)
Hupfer, C., Käpylä, P., Stix, M.: Reynolds stresses - dependence on latitude. Astronomische Nachrichten **326**(3–4), 223–226 (2005)
Käpylä, P.J.: Local numerical modelling of magnetoconvection and turbulence - implications for mean-field theories. PhD thesis, University of Helsinki, Finland, January (2006)
Meunier, N., Nesme-Ribes, E., Collin, B.: Cycle 19 facula dynamics. II. Meridional circulation, RMS velocity, and Reynolds stresses. Astron. Astrophys. **319**, 683–693 (1997)
Miesch, M.S.: Large-scale dynamics of the convection zone and tachocline. Living Rev. Solar Phys. **2**(1), 1 (2005)
Mulec, M.: Coronal Bright Points as Tracers for Solar Differential Rotation and Their Evolution. Diplomarbeit, Karl-Franzens University of Graz (2007)
Nash, A.G., Jr. Sheeley, N.R., Wang, Y.M.: Mechanisms for the rigid rotation of coronal holes. Solar Phys. **117**(2), 359–389 (1988)
Ossendrijver, M.: The solar dynamo. Astron. Astrophys. Rev. **11**, 287–367 (2003)
Pulkkinen, P., Tuominen, I.: Velocity structures from sunspot statistics in cycles 10 to 22. II. Latitudinal velocity and correlation functions. Astron. Astrophys. **332**, 755–760 (1998)
Rüdiger, G., Hollerbach, R.: The Magnetic Universe : Geophysical and Astrophysical Dynamo Theory. Wiley-VCH-Varlag (2004)
Ruediger, G.: Differential rotation and stellar convection. Sun and the solar stars. Akademie-Verlag, Berlin (1989)
Ruediger, G.: Reynolds stresses and differential rotation. II - Mean-field models. Geophys. Astrophys. Fluid Dyn. **21**(1), 1–25 (1982)
Ruediger, G.: Reynolds stresses and differential rotation. I. On recent calculations of zonal fluxes in slowly rotating stars. Geophys. Astrophys. Fluid Dyn. **16**(1), 239–261 (1980)
Ruždjak, D., Brajša, R., Skokić, I., Sudar, D., Hanslmeier, A.: On the periodicities in solar rotation and activity determined from sunspot-group data and possible connection with Rossby-like waves. Solar Phys. **298**(3), 39 (2023)
Ruždjak, D., Sudar, D., Brajša, R., Skokić, I., Poljančić Beljan, I., Jurdana-Šepić, R., Hanslmeier, A., Veronig, A., Pötzi, W.: Meridional motions and Reynolds stress determined by using Kanzelhöhe drawings and white light solar images from 1964 to 2016. Solar Phys. **293**(4), 59 (2018)
Sanchez, S., Fournier, A., Aubert, J.: The predictability of advection-dominated flux-transport solar dynamo models. Astrophys. J. **781**(1), 8 (2014)
Schmidt, W.: Modellrechnungen zur differentiellen Rotation der Sonne, Dissertation, University of Freiburg, Germany (1981)
Schmidt, W.: Models of solar differential rotation. Geophys. Astrophys. Fluid Dyn. **21**(1), 27–57 (1982)
Schroeter, E.H., Woehl, H.: Differential rotation and giant cell circulation of solar Ca^+-network. Solar Phys. **49**(1), 19–32 (1976)
Stix, M.: The Sun's differential rotation. Rev. Mod. Astron. **2**, 248–266 (1989)
Stix, M.: The Sun: An Introduction. Springer, Berlin (2002)
Sudar, D., Brajša, R., Skokić, I., Poljančić Beljan, I., Wöhl, H.: Meridional motion and Reynolds stress from Debrecen photoheliographic data. Solar Phys. **292**(7), 86 (2017)
Sudar, D., Brajša, R., Skokić, I., Temmer, M., Saar, S.: Rotation profiles of coronal bright points inside and outside of coronal holes. Solar Phys. **299**, 50 (2024)
Sudar, D., Saar, S.H., Skokić, I., Poljančić Beljan, I., Brajša, R.: Meridional motions and Reynolds stress from SDO/AIA coronal bright points data. Astron. Astrophys. **587**, A29 (2016)
Sudar, D., Skokić, I., Ruždjak, D., Brajša, R., Wöhl, H.: Tracing sunspot groups to determine angular momentum transfer on the Sun. Mon. Not. **439**(3), 2377–2385 (2014)
Vršnak, B., Brajša, R., Wöhl, H., Ruždjak, V., Clette, F., Hochedez, J.F.: Properties of the solar velocity field indicated by motions of coronal bright points. Astron. Astrophys. **404**, 1117–1127 (2003)

Wang, Y.M., Jr. Sheeley, N.R.: Understanding the rotation of coronal holes. Astrophys. J. **414**, 916 (1993)

Ward, F.: The general circulation of the solar atmosphere and the maintenance of the equatorial acceleration. Astrophys. J. **141**, 534 (1965)

Wöhl, H., Brajša, R.: Meridional motions of stable recurrent sunspot groups. Solar Phys. **198**(1), 57–77 (2001)

Chapter 8
Solar Rotation and Activity

The topic of this last chapter of the book is the complex relation between solar rotation and solar activity. First, representative results of observations will be presented: solar rotation before and during the Maunder minimum, cycle-related variations of the solar rotation from the nineteenth century till present, and indications of the north–south asymmetry of the solar rotation. Then, possible explanations of these observational results will be offered, including sampling and statistical influences and an interplay between the Reynolds and Maxwell stresses. An alternative method for analysis of the solar rotation variations will be described.

8.1 Results of Observations

Observational results will be presented separately for the following cases:

- behavior of the solar rotation before the Maunder minimum;
- behavior of the solar rotation during the Maunder minimum;
- variations of the solar rotation with the solar activity cycle;
- indications of the north–south asymmetry of the solar rotation.

8.1.1 Solar Rotation Before the Maunder Minimum

The Maunder minimum was a period in the years 1645–1715 when solar activity was significantly reduced and sunspots were almost completely absent on the Sun. We address the topic in the next subsection. It is important to note that after the first telescopic observations of sunspots in the early seventeenth century and the beginning of the Maunder minimum in 1645 there were not many reliable observations of sunspot positions. One notable exception is the observations of Christoph Scheiner

R. Brajša and A. Hanslmeier, *Solar Rotation*, UNITEXT for Physics,
https://doi.org/10.1007/978-981-97-6879-0_8

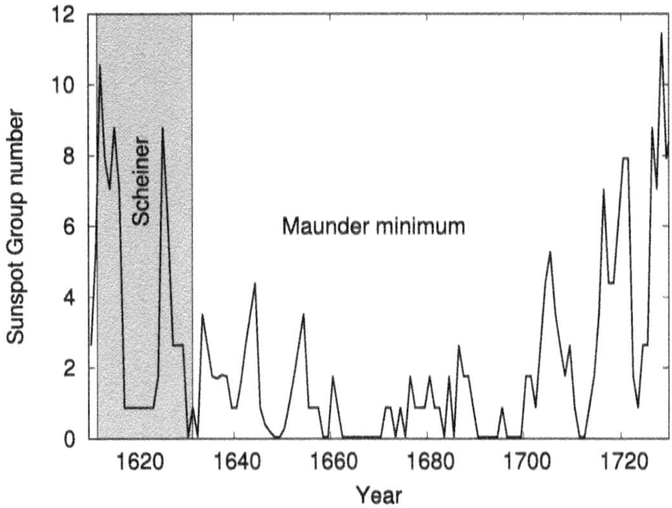

Fig. 8.1 Christoph Scheiner observed sunspots in the period 1611–1631 (denoted with the gray area in the panel) before the Maunder minimum. The solid line represents solar activity using the Sunspot Group Number according to Svalgaard and Schatten (2016). Credit: Sudar and Brajša (2022)

performed in the years 1611–1631. The sunspot positions determined from observations of Ch. Scheiner became available only recently (Arlt et al. 2016) and were subsequently used for the determination of the solar rotation and related phenomena (Sudar and Brajša 2022). Sunspot observations of Ch. Scheiner just precede the Maunder minimum, as can be seen in Fig. 8.1.

The differential rotation profile determined by Scheiner's data is presented in Fig. 6.3 and the differential rotation parameters are given in Table 6.3, together with later results, including modern measurements. It is interesting that in the years 1611–1631 the equatorial rotation velocity, given with the parameter A, is lower than later measurements (after the Maunder minimum). Further, the rotation profile is more differential (given by the absolute value of the parameter B) than any other result. This means that Scheiner's differential rotation curve is systematically lower than other measurements, indicating average lower rotation velocities at all latitudes. This might represent an important piece of evidence about the behavior of solar rotation immediately before the Maunder minimum.

In Fig. 8.2 meridional motions determined using Scheiner's data are presented. The qualitative behavior is similar to the profile of meridional flow determined by tracing sunspot groups after the Maunder minimum (e.g., Fig. 7.1). Scheiner's data also show poleward motions at low latitudes and equatorward motions at medium latitudes, but the amplitudes of the mean meridional velocities are larger (see Fig. 8.2). This could also represent another piece of observational evidence about the behavior of the solar rotation in the time just before the Maunder minimum.

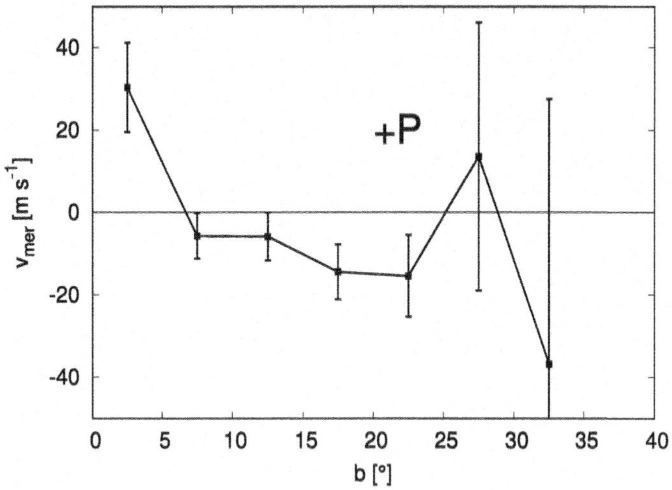

Fig. 8.2 Meridional motions of sunspot groups as a function of the latitude determined using sunspot observations of Ch. Scheiner in the years 1611–1631. Positive velocity values refer to the motions towards the pole and negative velocity values indicate motions towards the equator. Average values for the 5 deg latitude bins are given with the error bars. Both solar hemispheres were treated together. Credit: Sudar and Brajša (2022)

Finally, we present the correlation of velocity components and the latitudinal dependence of the horizontal component of Reynolds stress for the Scheiner data set in Figs. 8.3 and 8.4, respectively. For the correlation of the velocity components the following relation was found:

$$v_{\text{mer}} = (-0.023 \pm 0.011)\,\Delta v_{\text{rot}} + (-6.6 \pm 3.1)\,\text{ms}^{-1}\ . \tag{8.1}$$

Probably the most important property of Scheiner's data is the profile of the latitudinal dependence of the horizontal Reynolds stress, presented in Fig. 8.4. It can clearly be seen that the minimum of the best-fit curve is shifted to lower latitudes (less than 10 deg) than for the modern measurements (at about 40 deg), while the amplitudes are similar for both cases. This might represent an early indication of the Maunder-type minima and can be possibly used as a precursor for the forecast of the future grand minima of this type, as discussed in detail in Sudar and Brajša (2022). For completeness, we present also the best-fit parameters for the curve from Fig. 8.4 in Table 8.1, using Eq. (7.25) repeated here:

$$Q = c_1\, b\, \mathrm{e}^{-c_3\, b^2}\ , \tag{8.2}$$

where Q is the horizontal Reynolds stress, b is the latitude, and c_1 and c_3 are coefficients to be determined by fitting results of measurements. We note that in the paper Sudar and Brajša (2022) there is a typing error in the sign of the parameter c_1; it

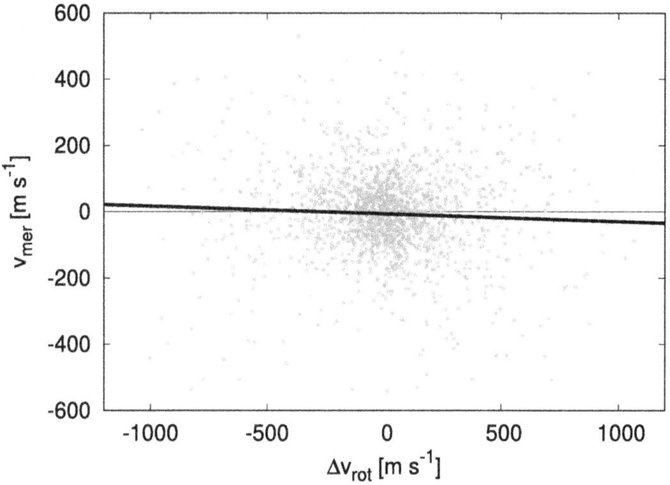

Fig. 8.3 Correlation of velocity components, meridional velocity (v_{mer}) and rotation velocity residual (Δv_{rot}), determined by tracing sunspot groups in the period 1611–1631 using observational data of Ch. Scheiner. Positive meridional velocities refer to the motions towards the pole and negative meridional velocities refer to the motions towards the equator. Positive rotation velocity residuals denote faster rotation than average and negative rotation velocity residuals denote slower rotation than average. The linear fit is denoted with the full line and both solar hemispheres are treated together. Credit: Sudar and Brajša (2022)

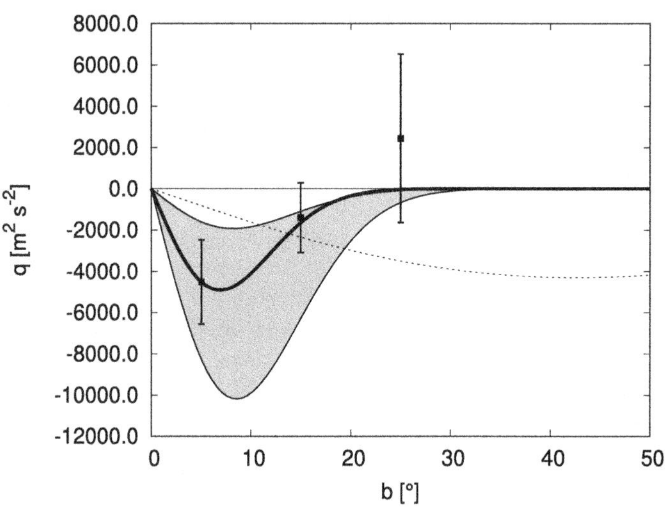

Fig. 8.4 The Reynolds stress values (q) averaged in the bins of 10 deg of latitude (b) are shown with black squares together with corresponding error bars, calculated using the observational data of Ch. Scheiner from the years 1611–1631. The thick black line shows the best-fit curve, and the gray area shows the uncertainty region defined by the errors of the fitting coefficients. The best-fit curve for modern-day observations (Sudar et al. 2017) is shown as a dotted line. Credit: Sudar and Brajša (2022)

Table 8.1 The best-fit coefficients from Eq. (7.25), repeated here as Eq. (8.2), for the Scheiner's data set (Sudar and Brajša 2022)

Coefficient	Value	Relative error (%)
c_1 [$m^2\ s^{-2}\ deg^{-1}$]	-1176 ± 790	67
c_3 [deg^{-2}]	0.011 ± 0.007	64

should be negative, since the required transport of angular momentum is towards the solar equator.

8.1.2 Solar Rotation During the Maunder Minimum

The Maunder minimum was a prolonged period of substantially reduced solar activity with very few sunspots observed on the Sun. The 11-year solar activity was still present, but barely detectable, according to the data reconstructed later. The Maunder minimum lasted from 1645 to 1715 and it partly coincided with the so-called Little Ice Age, when Europe and North America had lower temperatures than average. More details about the Maunder minimum can be found in Eddy (1976) and Hanslmeier (2020).

The main question is whether the solar rotation changed during the Maunder minimum, as compared with the periods of "normal" solar activity, before and after it. Answer to this question is not easy because of rather sparse observations, due to the lack of sunspots to be observed. However, several successful analyses were made to reconstruct the behavior of solar rotation during the Maunder minimum and we summarize here a hopefully reliable, but only a rather overall picture.

So, a higher rotation velocity at low latitudes during the Maunder minimum was reported in Eddy et al. (1976), based on the observations by Hevelius. Further, a more differential rotation profile during the Maunder minimum, in comparison to the modern values, was reported by Abarbanell and Wöhl (1981) and Ribes and Nesme-Ribes (1993). The solar rotation velocity during the Maunder minimum was also analyzed in Mendoza (1999), where it was additionally suggested the existence of a positive correlation between the solar rotation velocity and the solar cycle length for cycles nos. 12–20. A faster rotation velocity during the Maunder minimum is consistent with the estimated solar cycle length of about 17 years.

8.1.3 Cycle-Related Variations of the Solar Rotation

There is huge literature about the variations of the solar rotation related to the activity cycle. The majority of research papers are based on sunspot observations, since astronomers have more than 400 years of telescopic sunspot observations at disposal, but we stress that variation of the solar rotation was studied also with other tracers and methods. In many papers dealing with sunspot rotation, already cited in this book, also the temporal variation of the solar rotation velocity is analyzed. For an overview of the topic we refer the reader to the following papers, each from the one of the last four decades, Brajša et al. (1997, 2006), Ruždjak et al. (2017), Poljančić Beljan et al. (2022), and references therein.

One of the ways to follow possible variations of the solar rotation is to search for temporal changes in the solar rotation parameters from Eq. (6.1). In the case of sunspots, the solar differential parameter C is mostly set to zero, since there are no sunspots at high solar latitudes and in polar regions. This makes the analysis simpler, because the cross-talk between the parameters B and C is avoided.

We begin with the analysis of temporal and cycle-related changes in the solar differential rotation determined by tracing sunspots and sunspot groups in the period 1880–2016 (Ruždjak et al. 2017). Secular changes of the solar equatorial rotation velocity, expressed through the rotation parameter A, are presented in Fig. 8.5. A gradual decrease of the equatorial rotation velocity was found, coincidentally with the gradual increase of solar activity expressed through the mean yearly total sunspot number. In Ruždjak et al. (2017) various data sets, partly overlapping in time, were analyzed individually and combined. The anticorrelation between the solar rotation parameter A and solar activity was found in all analyzed cases, but the behavior of the parameter B is more complicated. In some cases the obtained results are in

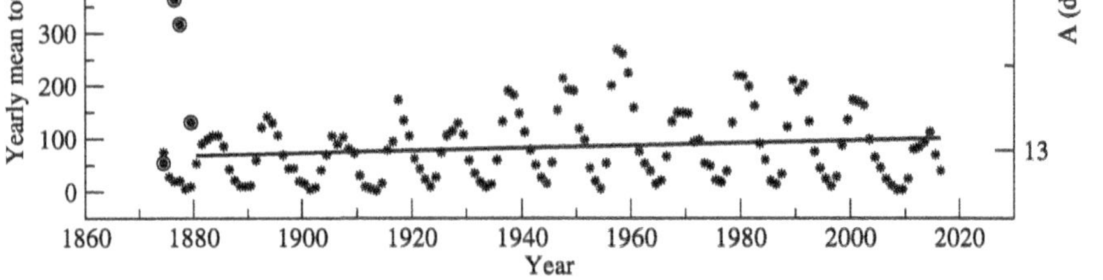

Fig. 8.5 Secular variations of the solar rotation and activity for the period 1880–2016. The mean yearly values for both quantities are given in the figure. Solar equatorial rotation velocity, expressed with the solar rotation parameter A, (denoted by black circles) was determined by tracing sunspot groups using positions from the Greenwich Photoheliographic Results (GPR) and Debrecen Photoheliographic Data (DPD). The yearly mean total sunspot numbers (denoted by asterisks) were taken from WDC-SILSO catalogue prepared by the Royal Observatory of Belgium, Brussels, recalibrated version 2.0. The lines represent linear fits of the data. The circled points denote the suspiciously low values of the equatorial rotation velocity, recorded at the beginning and the end of the considered time interval, excluded from the further analysis. Credit: Ruždjak et al. (2017)

mutual agreement, for the same time periods, but in the other significant differences were found. A nice example is presented in Figs. 8.6 and 8.7 for the period 1977–2016 (Ruždjak et al. 2017). The USAF/NOAA data set was used in that analysis. A negative correlation was found between the solar rotation parameter A and the solar activity (Fig. 8.6) and a positive correlation between the absolute value of the solar rotation parameter B and the solar activity (Fig. 8.7). These results were found to be partly statistically significant; further details can be found in Ruždjak et al. (2017).

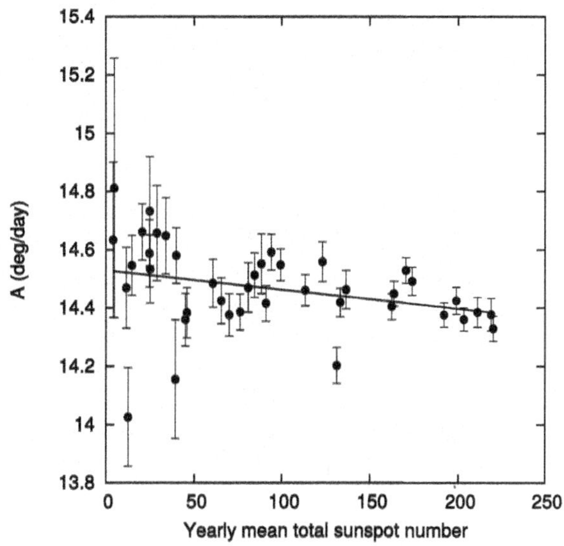

Fig. 8.6 The yearly values of the solar equatorial rotation velocity, expressed through the rotation parameter A, as a function of the yearly mean total sunspot number. Solar rotation velocity was determined by tracing sunspot groups from the USAF/NOAA data set in the period 1977–2016. The line represents the least-squares fit using the Marquardt–Levenberg algorithm. Credit: Ruždjak et al. (2017)

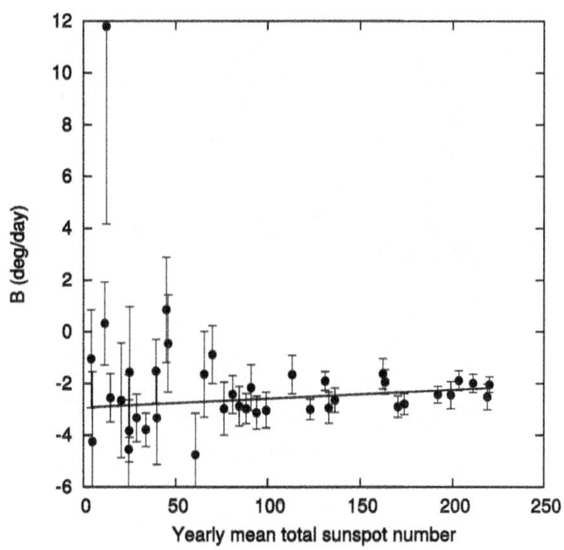

Fig. 8.7 Similar to Fig. 8.6, but for the solar differential rotation parameter B. Credit: Ruždjak et al. (2017)

However, the above-mentioned results regarding a dependence of the solar rotation parameters on the level of solar activity could not have been confirmed using all available sunspot data sets. One notable example is the analysis of the variation of the solar differential rotation and activity in the period 1964–2016 using the Kanzelhöhe data set (Poljančić Beljan et al. 2022). The observing period of this analysis embraces the years of the previous study (USAF/NOAA data) and extends further in the past. The results obtained using the Kanzelhöhe data set indicate qualitatively a completely different behavior. A positive correlation between the solar rotation parameter A and the sunspot number was found as well as a negative correlation between the solar rotation parameter B and the sunspot number (see Fig. 4 in Poljančić Beljan et al. (2022)). We should note, however, that these results are not statistically significant.

As we have seen, the results on the dependence of the solar rotation on the activity level using sunspots are partly contradictory and are not completely conclusive. So, we search for additional data to provide hopefully more insight into this controversial topic. A dependence of the solar rotation determined by tracing coronal bright points in 28.4 nm SOHO/EIT solar images in the period 1998–2006 is presented for the differential rotation parameters A and B, in Figs. 8.8 and 8.9, respectively (Jurdana-Šepić et al. 2011). A clear negative and highly significant correlation between the solar rotation parameter A and the monthly smoothed relative sunspot number was found (Fig. 8.8). On the other hand, the result for the solar differential rotation parameter B is consistent with no correlation between this parameter and the solar activity level expressed through the monthly smoothed relative sunspot number

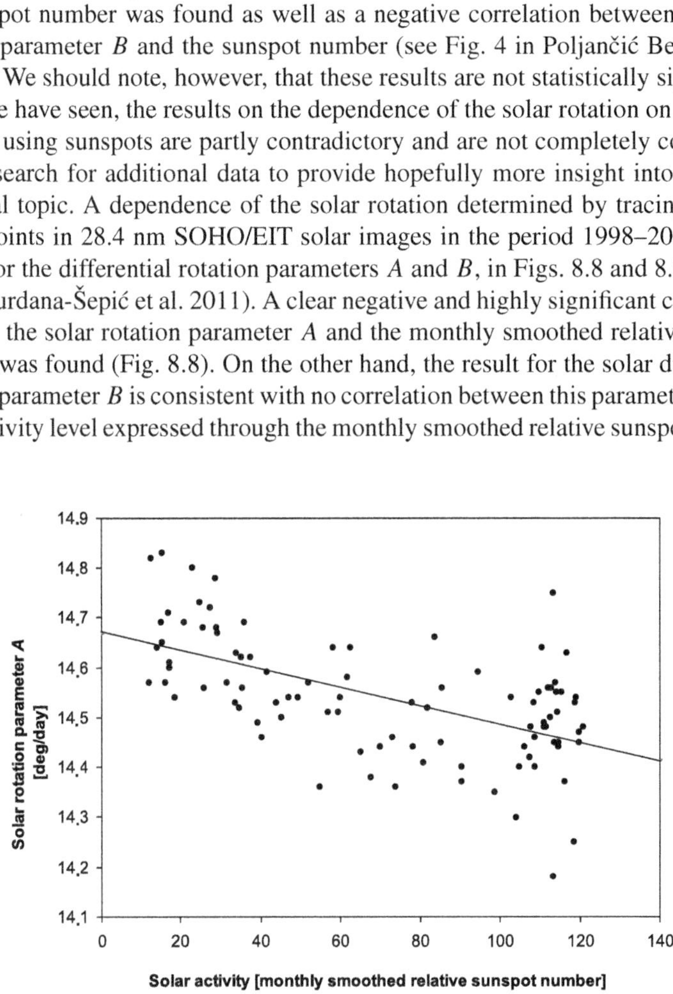

Fig. 8.8 Monthly values of the solar rotation parameter A as a function of the monthly smoothed relative sunspot number taken from WDC-SILSO catalogue prepared by the Royal Observatory of Belgium, Brussels, version 1.0. Solar rotation was determined by tracing coronal bright points in the 28.4 nm SOHO/EIT full-disk solar images taken in the years 1998–2006. In total 93 monthly values of the solar rotation parameters were determined and used in the analysis. Adapted from: Jurdana-Šepić et al. (2011)

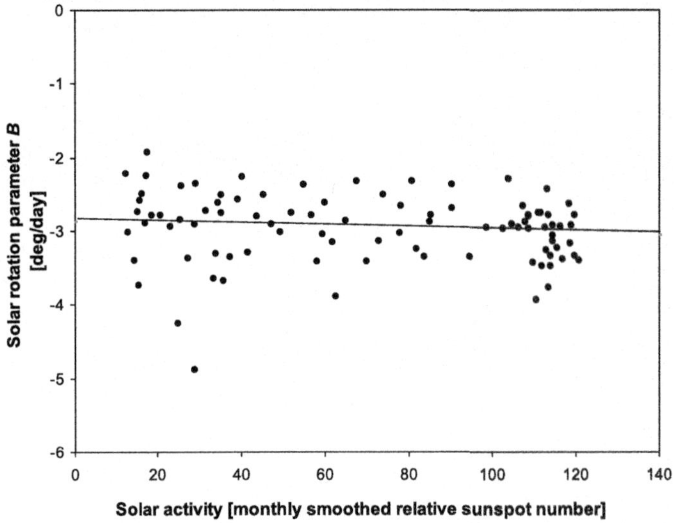

Fig. 8.9 Similar to Fig. 8.8, but for the solar differential rotation parameter B. Adapted from: Jurdana-Šepić et al. (2011)

(Fig. 8.9). One of the aims of that study (Jurdana-Šepić et al. 2011) was also the extend the analysis to the heliospheric data. A relationship between the solar rotation and the interplanetary magnetic field derived from the interdiurnal variability (IDV) index (Svalgaard and Cliver 2005) is presented for the solar rotation parameters A and B in Figs. 8.10 and 8.11, respectively. Note that in this part of analysis the yearly values are used, instead of the monthly values. Similar to the previous case, a highly statistically significant negative correlation between the solar rotation parameter A and the interplanetary magnetic field was found, but the relationship of the differential rotation parameter B and the interplanetary magnetic field is consistent with no correlation.

8.1.4 North–South Asymmetry of the Solar Rotation

In the study of solar rotation and related phenomena the question of symmetry has an important role. In the previous chapter we have already encountered examples of non-axisymmetric meridional motions. So, the symmetry around the axis of solar rotation is broken for some observed phenomena. Another interesting symmetry is the north–south symmetry of the solar rotation velocity. This symmetry is respective to the solar equator. It is important to note that in many representations of the solar rotation it is assumed that solar rotation is symmetric around the solar equator. This is the case for reduction of measured data, as well as in many theoretical modeling efforts.

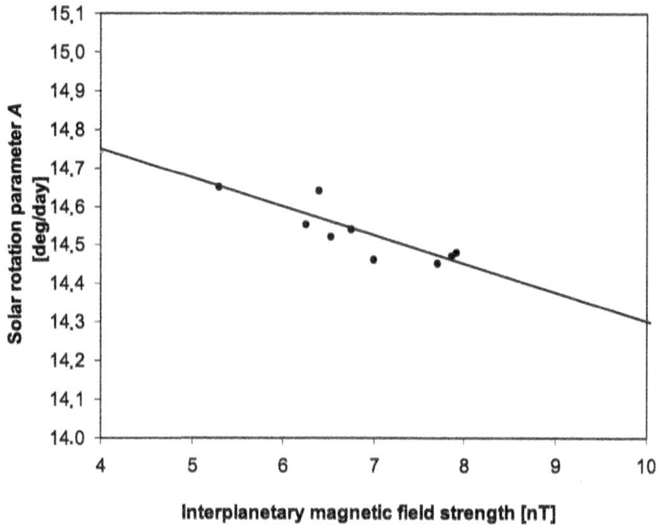

Fig. 8.10 Yearly values of the solar rotation parameter A as a function of the interplanetary magnetic field derived from the interdiurnal variability (IDV) index. Credit: Jurdana-Šepić et al. (2011)

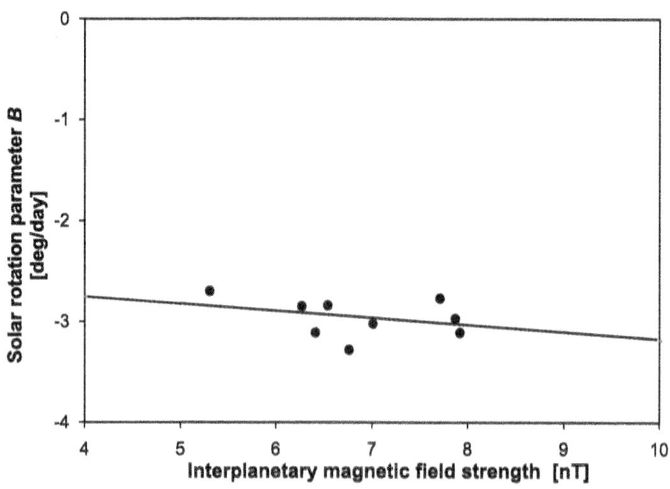

Fig. 8.11 Similar to Fig. 8.10, but for the solar differential rotation parameter B. Credit: Jurdana-Šepić et al. (2011)

There are many observational indications that solar rotation is not symmetric respectively to the solar equator. Here we give a couple of observational examples obtained using various tracers. A north–south asymmetry of the solar rotation was reported for Hα filaments (Brajša et al. 1997), for low brightness temperature regions observed at mm wavelengths (Brajša et al. 1997, 2000), for coronal bright points (Brajša et al. 2002; Wöhl et al. 2010), and for sunspots and sunspot groups (Brajša et al. 2006; Poljančić Beljan et al. 2017). Further examples can be found in the references therein.

We present here two examples of the north–south asymmetry of the solar rotation determined by tracing coronal bright points in the 28.4 nm SOHO/EIT full-disk solar images (Wöhl et al. 2010). Two methods for rotation velocity determination (an interactive method and an automatic method) were applied for data recorded in two time periods: 1998–1999 and 1998–2006. Further details about the two mentioned data reduction methods are given in Wöhl et al. (2010) and references therein. The average rotation velocity values and the corresponding differential rotation profiles are presented in Figs. 8.12 and 8.13. Now we focus on the differential character of the solar rotation profile. For both time intervals, 1998–1999 and 1998–2006, the southern solar hemisphere rotated less differential, i.e., showed a rigid component at middle and high latitudes. The differential character of the solar rotation is expressed through the solar rotation parameter B. The difference between the solar rotation parameter B for the northern and the southern solar hemisphere is statistically significant on the 1σ level (data from 1998–1999) and on the 2σ level (data from 1998–2006), and further details with numerical values with errors can be found in Wöhl et al. (2010).

Besides checking whether the north–south asymmetry of the solar rotation is statistically significant, another important question is if a pattern can be revealed which could provide some insight into why there exists a north–south rotational asymmetry and how it changes with time. For the coronal bright points data presented here (Figs. 8.12 and 8.13) such pattern was indeed found. During the ascending phase of the solar cycle no. 23 (1998–1999) and during most of that solar cycle, when coronal bright points were traced (1998–2006) the southern solar hemisphere was more active (Temmer et al. 2006, Fig. 6 and Table 2 in that paper). So, a stronger magnetic activity in the southern solar hemisphere was measured during the same time period when a less differential rotation curve of the same, southern solar hemisphere was observed.

8.2 Possible Explanations of the Observational Results

First we briefly summarize the most important observational results regarding interdependence of the solar rotation and activity:

- Immediately before the Maunder minimum the solar rotation curve was systematically shifted to lower velocity values in comparison to the period after the Maunder minimum.
- A possible precursor of the Maunder minimum might be the observed shift towards lower latitudes of the minimum in the Reynolds stress profile as a function of the latitude.

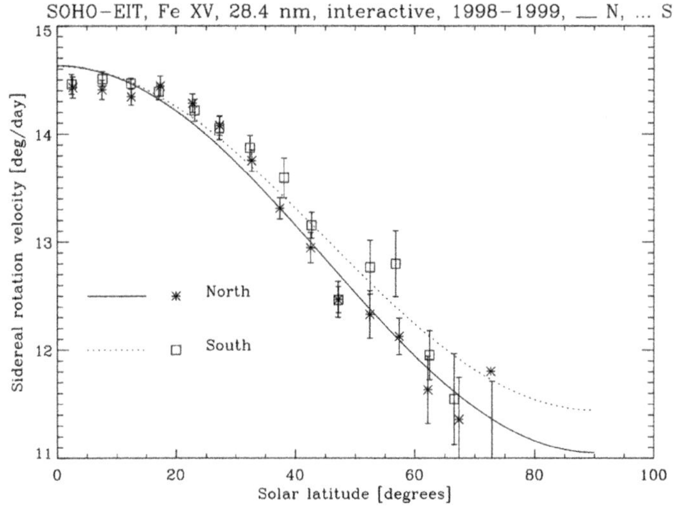

Fig. 8.12 The north–south asymmetry of the solar rotation for the period 1998–1999 determined by tracing coronal bright points in the 28.4 nm SOHO/EIT images. The mean values for 5 deg latitude bins with error bars are given together with the differential rotation curves. Results for the two solar hemispheres are presented with different symbols and lines, as denoted in the legend. Credit: Wöhl et al. (2010)

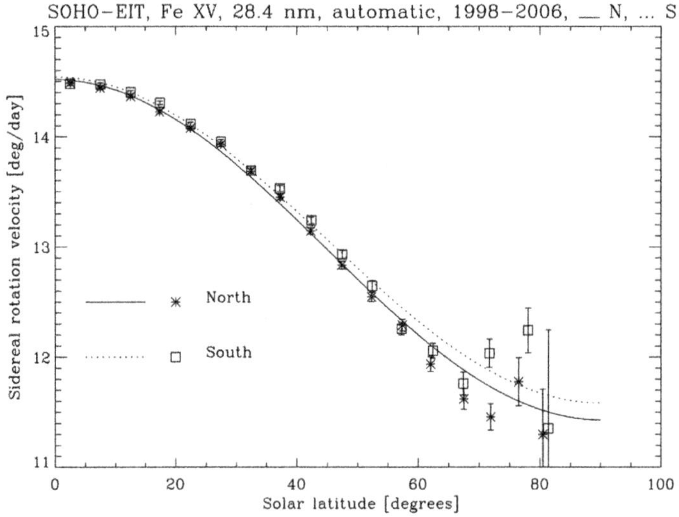

Fig. 8.13 Similar to Fig. 8.12 for the years 1998–2006. Credit: Wöhl et al. (2010)

- During the Maunder minimum a higher rotation velocity at low solar latitudes and a more differential rotation profile was found.
- Most observations suggest a clear relationship between solar equatorial rotation velocity and the activity level, both on the secular scale and in individual solar cycles: a lower solar activity coincides with a higher equatorial rotation velocity (the parameter A) and vice versa; an indication of a similar behavior was found also for the interplanetary magnetic field.
- Some observations indicate a more differential rotation profile (larger absolute values of the parameter B) for the years of lower solar activity, but these results are still not fully conclusive.
- Regarding the north–south rotational asymmetry, indications were found that a more active solar hemisphere shows a less differential solar rotation.

8.2.1 Sampling and Statistical Influences

Sunspots are widely used as tracers for the solar rotation determination with the aim to study a complex relationship between solar rotation and activity. The main advantages of this approach are the long data series of both sunspot positions and proxies for the solar activity, e.g., the relative sunspot number. However, there are quite a few disadvantages of using sunspots for this research, the most important being the following:

- Sunspots appear only at middle and low solar latitudes; they are rarely seen in equatorial regions and are completely absent at high latitudes and in polar regions.
- During the solar cycle sunspots migrate from middle to lower solar latitudes, which is nicely presented in the butterfly diagram; so, in any phase of the solar cycle sunspots occupy only a subset of the complete sunspot latitudinal range.
- There is a huge difference in the number of sunspots present on the Sun between the maximum and minimum of the 11-year solar cycle.
- Solar cycles often overlap: as a result, in a small fraction of time sunspots from the old and from the new solar cycle might be present on the Sun simultaneously.

Taking all this into account a serious question arises, namely whether the cycle-related changes of the solar rotation determined by tracing sunspots are of the solar origin, or appear as a result of the above-mentioned properties.

We propose two solutions for the above-mentioned problem. The first possibility is to use another tracer, e.g., coronal bright points. We have already seen many good examples of using coronal bright points for the determination of the solar rotation and related phenomena. The EUV data series are much shorter than the sunspot data series, but some satellite EUV data sets (e.g., from SOHO and SDO) are rather promising, having high spatial and temporal resolutions and in some cases already covering at least one whole solar cycle.

The second possibility is to find a different data reduction procedure, thinking a bit "out of the box". An innovative approach to the problem, a so-called residual method, was introduced and successfully applied to the Mount Wilson Doppler and sunspot data in Gilman and Howard (1984). This method was later applied also to other data sets (see, e.g., Brajša et al. (1997, 2004, 2006)).

In the residual method the yearly deviations $\Delta\omega(b, i)$ from the mean rotation velocity, averaged over all years under consideration, are calculated for each latitude band:

$$\Delta\omega(b, i) = \hat{\omega}(b, i) - \hat{\omega}(b, c) \ , \tag{8.3}$$

where $\hat{\omega}(b, i)$ is the averaged rotation velocity for the year i and for a latitude band centered at the latitude b, and $\hat{\omega}(b, c)$ is given by

$$\hat{\omega}(b, c) = \frac{\sum_{i=1}^{n} \hat{\omega}(b, i)}{n} \ . \tag{8.4}$$

The total number of years under consideration is denoted with n. These deviations are then averaged over latitudes and the yearly rotation velocity residuals $\widehat{\Delta\omega}(i)$ are calculated providing a single number for each year:

$$\boxed{\widehat{\Delta\omega}(i) = \frac{\sum_{b=0}^{90} \Delta\omega(b, i)\, m(b, i)}{P(i)}} \ , \tag{8.5}$$

where $m(b, i)$ is the number of rotation velocities in a latitude band centered at the latitude b in the year i, and $P(i)$ is the sum of $m(b, i)$ over latitudes:

$$P(i) = \sum_{b=0}^{90} m(b, i) \ . \tag{8.6}$$

A positive rotation velocity residual indicates a higher than average velocity and a negative rotation velocity residual indicates a lower than average velocity.

Secular variation of the rotation velocity residual for the extended Greenwich data set is presented in Fig. 8.14 (Brajša et al. 2006). A secular decrease of the rotation velocity residual was found to be in a general agreement with the secular decrease of the equatorial rotation velocity shown in Fig. 8.5. In these two cases different combinations of sunspot data sets, partly overlapping, were used and the time intervals were mostly the same. Finally, it is noteworthy to point out that three different data sets (the extended Greenwich sunspot data set, the Mount Wilson sunspot data set, and the Mount Wilson Doppler data set) showed a peak in the rotation velocity residual around solar activity minima (Gilman and Howard 1984; Brajša et al. 2006).

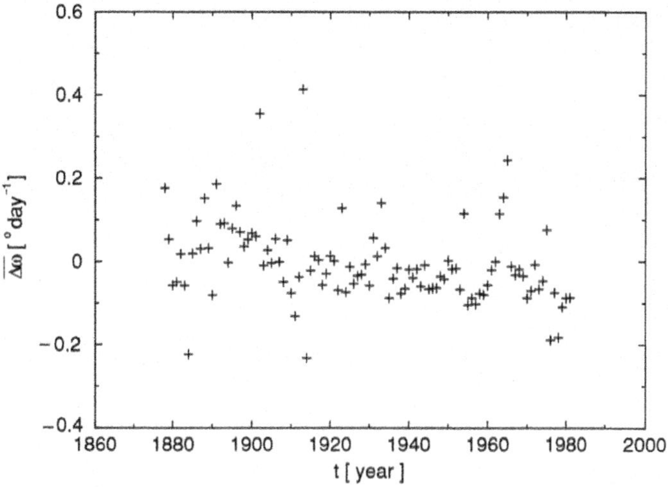

Fig. 8.14 Rotation velocity residual as a function of time for the years 1878–1981. The extended Greenwich data set was used for the analysis. Credit: Brajša et al. (2006)

8.2.2 Reynolds Stress Versus Maxwell Stress

In this chapter, we have presented observational evidence for two main findings regarding the relationship between solar rotation and activity. First, a higher equatorial and low latitude rotation velocity was found during low solar activity and the rotation velocity decreases with increasing activity level. Further, solar rotation is more differential during times of low activity and becomes less differential with an increase in activity. The second result refers to the north–south asymmetry of the solar rotation. A more active solar hemisphere shows a less differential rotation profile.

We explain both of these results with the interplay between the Reynolds stress and the Maxwell stress and recall Eq. (7.5) where the components of the angular momentum flux are given:

$$\boxed{F_i = r\sin\theta\,\rho\left\{\underbrace{\langle v_i\rangle\,\omega\,r\sin\theta}_{\text{Meridional flow}} + \underbrace{\langle v_i'\,v_\phi'\rangle}_{\text{Reynolds stress}} - \frac{1}{4\pi\rho}\underbrace{(\langle B_i'\,B_\phi'\rangle + \langle B_i\rangle\langle B_\phi\rangle)}_{\text{Maxwell stress}}\right\}} \tag{8.7}$$

It can be seen that the Maxwell stress opposes the Reynolds stress reducing the angular momentum transport towards the equator. In this way the differential character of the solar rotation is also reduced. The Maxwell stresses are associated with the correlations of the fluctuating or turbulent, non-axisymmetric magnetic field components which arise from tilt and twist within magnetic structures (Brun et al. 2004).

They have the origin in the reaction on the flow by the Lorentz forces (e.g., Sturrock (1994), Parker (1996)).

According to Brajša et al. (2006) we propose the following explanation of the observed results. The interaction between differential rotation and magnetic fields in the solar convection zone was discussed in Brun (2004), where the three-dimensional numerical simulations of compressible convection under the influence of rotation and magnetic fields in spherical shells were described. One consequence of the model is that the Maxwell stresses can oppose the Reynolds stresses and, in this way, contribute to the transport of the angular momentum towards the solar poles, leading to a reduced differential rotation. So, when magnetic fields are weaker, one can expect a more pronounced differential rotation yielding a higher rotation velocity at low latitudes on an average. This is consistent with observational results of the solar rotation behavior during the Maunder minimum, during the secular variation of the solar rotation and for the individual solar activity cycles. Models of the angular momentum transport by magnetoconvection based on the interplay between the Reynolds stress and the Maxwell stress were further developed in Lanza (2007) and Rüdiger and Küker (2021) with the aim to explain variation of the solar differential rotation by magnetic field modulation.

Finally, we follow the same line of reasoning to explain the observed north–south rotational asymmetry (e.g., Wöhl et al. (2010)). Similar to the cycle-related variation of the solar rotation, here is the proposed explanation also within the model of the relationship between solar rotation and activity. The balance between the Reynolds stress and the Maxwell stress is considered to explain the observed less differential rotation of the southern solar hemisphere, in comparison to the northern one, during most of the 23rd solar cycle, when the southern hemisphere was more active. Solar differential rotation was to some extent suppressed in the southern hemisphere due to the stronger magnetic activity in this hemisphere at the time considered. Maxwell stresses, related to the magnetic field, oppose the usual Reynolds stresses, which are the main cause of the differential character of the solar rotation.

Exercises and Questions

1. What is the Maunder minimum and when it took place?

2. What was the main characteristic of the solar rotation before the Maunder minimum? Is it possible to identify a possible precursor of the Maunder-type grand solar minimum based on available observational evidence?

3. How the solar rotation looked like during the Maunder minimum?

4. Describe the temporal variation of the solar rotation and a complex relationship between the solar rotation and activity, based on various tracers and methods.

5. What is the relationship between the solar rotation parameters and the interplanetary magnetic field?

6. Describe observational evidence for the north–south asymmetry of the solar rotation.

7. What is the residual method for the reduction of the solar rotation data and how is it applied to the analysis of the temporal variation of the solar rotation?

8. How can the interplay between the Maxwell stress and the Reynolds stress explain the observed relationship between the solar rotation and activity? Consider the temporal, cycle-related variation and the north–south asymmetry of the solar rotation.

References

Abarbanell, C., Wöhl, H.: Solar rotation velocity as determined from sunspot drawings of Hevelius, J. in the 17th-century. Solar Phys. **70**(1), 197–203 (1981)

Arlt, R., Senthamizh Pavai, V., Schmiel, C., Spada, F.: Sunspot positions, areas, and group tilt angles for 1611-1631 from observations by Christoph Scheiner. Astron. Astrophys. **595**, A104 (2016)

Brajša, R., Ruždjak, V., Vršnak, B., Pohjolainen, S., Urpo, S., Schroll, A., Wöhl, H.: On the possible changes of the solar differential rotation during the activity cycle determined using microwave low-brightness regions and Hα filaments as tracers. Solar Phys. **171**(1), 1–34 (1997)

Brajša, R., Ruždjak, V., Vršnak, B., Wöhl, H., Pohjolainen, S., Upro, S.: Statistical weights and selective height corrections in the determination of the solar rotation velocity. Solar Phys. **196**(2), 279–297 (2000)

Brajša, R., Ruždjak, D., Wöhl, H.: Temporal variations of the solar rotation determined by sunspot groups. Solar Phys. **237**(2), 365–382 (2006)

Brajša, R., Wöhl, H., Ruždjak, D., Schawinski-Guiton, K.: Variation of the solar rotation during the activity cycle applying the residual method to Greenwich data. Hvar Obs. Bull. **28**(1), 55–62 (2004)

Brajša, R., Wöhl, H., Vršnak, B., Ruždjak, V., Clette, F., Hochedez, J.F.: Solar differential rotation determined by tracing coronal bright points in SOHO-EIT images. II. Results for 1998/99 obtained with interactive and automatic methods. Astron. Astrophys. **392**, 329–334 (2002)

Brun, A.S., Miesch, M.S., Toomre, J.: Global-scale turbulent convection and magnetic dynamo action in the solar envelope. Astrophys. J. **614**(2), 1073–1098 (2004)

Brun, A.S.: On the interaction between differential rotation and magnetic fields in the Sun. Solar Phys. **220**(2), 333–345 (2004)

Eddy, J.A., Gilman, P.A., Trotter, D.E.: Solar rotation during the Maunder minimum. Solar Phys. **46**, 3–14 (1976)

Eddy, J.A.: The Maunder minimum. Science **192**, 1189–1202 (1976)

Gilman, P.A., Howard, R.: Variations in solar rotation with the sunspot cycle. Astrophys. J. **283**, 385–391 (1984)

Hanslmeier, A.: The Chaotic Solar Cycle. Springer Nature, Singapore (2020)

Jurdana-Šepić, R., Brajša, R., Wöhl, H., Hanslmeier, A., Poljančić, I., Svalgaard, L., Gissot, S.F.: A relationship between the solar rotation and activity in the period 1998–2006 analysed by tracing small bright coronal structures in SOHO-EIT images. Astron. Astrophys. **534**, A17 (2011)

Lanza, A.F.: Angular momentum conservation and torsional oscillations in the Sun and solar-like stars. Astron. Astrophys. **471**(3), 1011–1022 (2007)

Mendoza, B.: Solar rotation and cycle length. Solar Phys. **188**(2), 237–243 (1999)

Parker, E.N.: The solar magnetic field. In: The Structure of the Sun, 6th Canary Islands Winter School of Astrophysics, p. 301. Cambridge University Press (1996)

Poljančić Beljan, I., Jurdana-Šepić, R., Brajša, R., Sudar, D., Ruždjak, D., Hržina, D., Pötzi, W., Hanslmeier, A., Veronig, A., Skokić, I., Wöhl, H.: Solar differential rotation in the period 1964-2016 determined by the Kanzelhöhe data set. Astron. Astrophys. **606**, A72 (2017)

Poljančić Beljan, I., Jurdana-Šepić, R., Jurkić, T., Brajša, R., Skokić, I., Sudar, D., Ruždjak, D., Hržina, D., Pötzi, W., Hanslmeier, A., Veronig, A.M.: Variation in solar differential rotation and activity in the period 1964-2016 determined by the Kanzelhöhe data set. Astron. Astrophys. **663**, A24 (2022)

Ribes, J.C., Nesme-Ribes, E.: The solar sunspot cycle in the Maunder minimum AD1645 to AD1715. Astron. Astrophys. **276**, 549 (1993)

Rüdiger, G., Küker, M.: Angular momentum transport by magnetoconvection and the magnetic modulation of the solar differential rotation. Astron. Astrophys. **649**, A173 (2021)

Ruždjak, D., Brajša, R., Sudar, D., Skokić, I., Poljančić Beljan, I.: A relationship between the solar rotation and activity analysed by tracing sunspot groups. Solar Phys. **292**(12), 179 (2017)

Sturrock, P.A.: Plasma Physics, An Introduction to the Theory of Astrophysical, Geophysical and Laboratory Plasmas. Cambridge University Press, Cambridge (1994)

Sudar, D., Brajša, R., Skokić, I., Poljančić Beljan, I., Wöhl, H.: Meridional motion and Reynolds stress from Debrecen photoheliographic data. Solar Phys. **292**(7), 86 (2017)

Sudar, D., Brajša, R.: Solar rotation in the period 1611–1631 determined using observations of Christoph Scheiner. Solar Phys. **297**(10), 132 (2022)

Svalgaard, L., Cliver, E.W.: The IDV index: its derivation and use in inferring long-term variations of the interplanetary magnetic field strength. J. Geophys. Res. (Space Phys.) **110**(A12), A12103 (2005)

Svalgaard, L., Schatten, K.H.: Reconstruction of the sunspot group number: the backbone method. Solar Phys. **291**(9–10), 2653–2684 (2016)

Temmer, M., Rybák, J., Bendík, P., Veronig, A., Vogler, F., Otruba, W., Pötzi, W., Hanslmeier, A.: Hemispheric sunspot numbers R_n and R_s from 1945–2004: catalogue and N-S asymmetry analysis for solar cycles 18–23. Astron. Astrophys. **447**(2), 735–743 (2006)

Wöhl, H., Brajša, R., Hanslmeier, A., Gissot, S.F.: A precise measurement of the solar differential rotation by tracing small bright coronal structures in SOHO-EIT images. Results and comparisons for the period 1998-2006. Astron. Astrophys. **520**, A29 (2010)

Index

R. Brajša and A. Hanslmeier, *Solar Rotation*, UNITEXT for Physics,
https://doi.org/10.1007/978-981-97-6879-0

SPRINGER NATURE

GPSR Compliance

The European Union's (EU) General Product Safety Regulation (GPSR) is a set of rules that requires consumer products to be safe and our obligations to ensure this.

If you have any concerns about our products, you can contact us on ProductSafety@springernature.com

In case Publisher is established outside the EU, the EU authorized representative is:

Springer Nature Customer Service Center GmbH
Europaplatz 3
69115 Heidelberg, Germany

The manufacturer's authorised representative in the EU is Springer Nature Customer Service Centre GmbH, Europaplatz 3, 69115 Heidelberg, Germany. If you have any concerns regarding our products, please contact ProductSafety@springernature.com

Printed and bound by CPI Group (UK) Ltd, Croydon, CR0 4YY
28/07/2026
02179292-0005